# STUDENT SOLUTIONS MANUAL

## CHAPTERS 1–19

# PHYSICS

FOR SCIENTISTS AND ENGINEERS SECOND EDITION

## A STRATEGIC APPROACH

Randall D. Knight

Pawan Kahol
*Missouri State University*

Donald Foster
*Wichita State University*

Larry Smith
*Snow College*

Scott Nutter
*Northern Kentucky University*

San Francisco Boston New York
Cape Town Hong Kong London Madrid Mexico City
Montreal Munich Paris Singapore Sydney Tokyo Toronto

Publisher: Adam Black, Ph.D.
Development Manager: Michael Gillespie
Development Editor: Alice Houston, Ph.D.
Project Editor: Martha Steele
Assistant Editor: Grace Joo
Media Producer: Deb Greco
Sr. Administrative Assistant: Cathy Glenn
Director of Marketing: Christy Lawrence
Executive Marketing Manager: Scott Dustan
Sr. Market Development Manager: Josh Frost
Market Development Associate: Jessica Lyons
Managing Editor: Corinne Benson
Sr. Production Supervisor: Nancy Tabor
Production Service: WestWords PMG
Illustrations: International Typesetting and Composition (ITC)
Cover Design: Yvo Riezebos Design and Seventeenth Street Studios
Manufacturing Manager: Evelyn Beaton
Manufacturing Buyer: Carol Melville
Text and Cover Printer: Bradford & Bigelow
Cover Image: Composite illustration by Yvo Riezebos Design; photo of spring by Bill Frymire/Masterfile



SBN-13: 978-0-321-51354-0
ISBN-10: 0-321-51354-1

www.aw-bc.com

2 3 4 5 6 7 8 9 10—B&B—10 09 08

# Contents

# Preface

This *Student Solutions Manual* is intended to provide you with examples of good problem-solving techniques and strategies. To achieve that, the solutions presented here attempt to:.

- Follow, in detail, the problem-solving strategies presented in the text.
- Articulate the reasoning that must be done before computation.
- Illustrate how to use drawings effectively.
- Demonstrate how to utilize graphs, ratios, units, and the many other "tactics" that must be successfully mastered and marshaled if a problem-solving strategy is to be effective.
- Show examples of assessing the reasonableness of a solution.
- Comment on the significance of a solution or on its relationship to other problems.

We recommend you try to solve each problem on your own before you read the solution. Simply reading solutions, without first struggling with the issues, has limited educational value.

As you work through each solution, make sure you understand how and why each step is taken. See if you can understand which aspects of the problem made this solution strategy appropriate. You will be successful on exams not by memorizing solutions to particular problems but by coming to recognize which kinds of problem-solving strategies go with which types of problems.

We have made every effort to be accurate and correct in these solutions. However, if you do find errors or ambiguities, we would be very grateful to hear from you. Please contact: knight@aw.com

**Acknowledgments for the First Edition**

We are grateful for many helpful comments from Susan Cable, Randall Knight, and Steve Stonebraker. We express appreciation to Susan Emerson, who typed the word-processing manuscript, for her diligence in interpreting our handwritten copy. Finally, we would like to acknowledge the support from the Addison Wesley staff in getting the work into a publishable state. Our special thanks to Liana Allday, Alice Houston, and Sue Kimber for their willingness and preparedness in providing needed help at all times.

Pawan Kahol
*Missouri State University*

Donald Foster
*Wichita State University*

**Acknowledgments for the Second Edition**

I would like to acknowledge the patient support of my wife, Holly, who knows what is important.

Larry Smith
*Snow College*

I would like to acknowledge the assistance and support of my wife, Alice Nutter, who helped type many problems and was patient while I worked weekends.

Scott Nutter
*Northern Kentucky University*

# CONCEPTS OF MOTION

# 1

**1.1. Solve:**

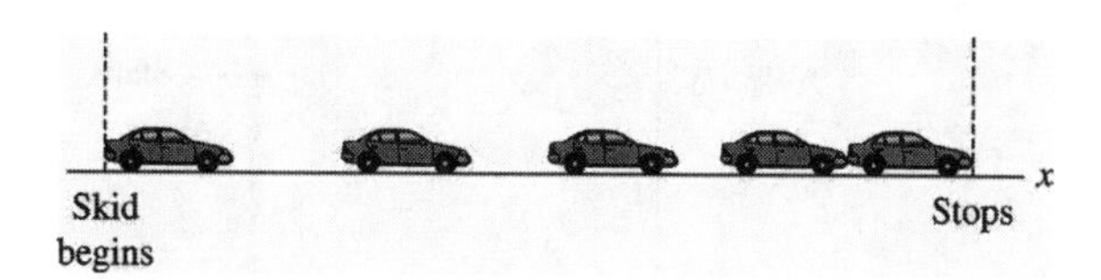


**1.9. Solve:** **(a)** Acceleration is found by the method of Tactics Box 1.3. Let $\vec{v}_0$ be the velocity vector between points 0 and 1 and $\vec{v}_1$ be the velocity vector between points 1 and 2.

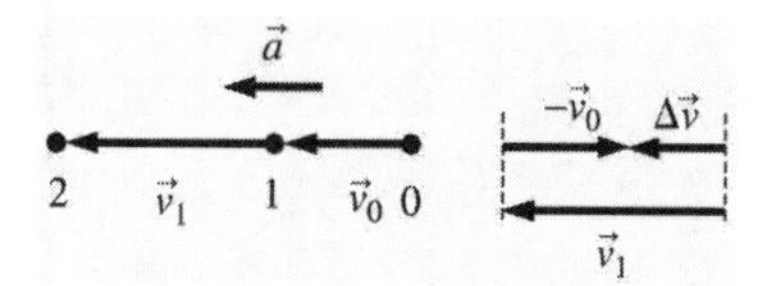


**(b)** Speed $v_1$ is greater than speed $v_0$ because $\vec{v}$ and $\vec{a}$ point in the same direction.

**1.13. Model:** Represent the car as a particle.

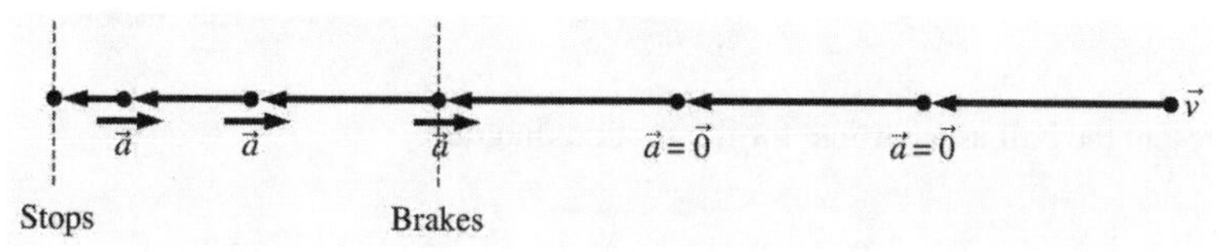


**Visualize:** The dots are equally spaced until brakes are applied to the car. Equidistant dots indicate constant average velocity. On braking, the dots get closer as the average speed decreases.

**1.21. Visualize:** The bicycle is moving with an acceleration of 1.5 m/s$^2$.

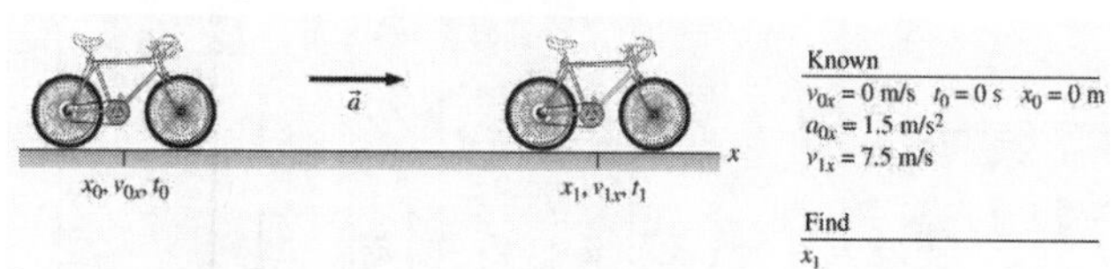


**1.23. Solve:** **(a)** $9.12\ \mu\text{s} = (9.12\ \mu\text{s})\left(\dfrac{10^{-6}\text{s}}{1\ \mu\text{s}}\right) = 9.12\times10^{-6}\ \text{s}$

**(b)** $3.42\ \text{km} = (3.42\ \text{km})\left(\dfrac{10^3\ \text{m}}{1\ \text{km}}\right) = 3.42\times10^3\ \text{m}$

**(c)** $44\ \text{cm/ms} = \left(44\dfrac{\text{cm}}{\text{ms}}\right)\left(\dfrac{10^{-2}\ \text{m}}{1\ \text{cm}}\right)\left(\dfrac{1\ \text{ms}}{10^{-3}\ \text{s}}\right) = 4.4\times10^2\ \text{m/s}$

**(d)** $80\ \text{km/hour} = \left(80\dfrac{\text{km}}{\text{hour}}\right)\left(\dfrac{10^3\ \text{m}}{1\ \text{km}}\right)\left(\dfrac{1\ \text{hour}}{3600\ \text{s}}\right) = 22\ \text{m/s}$

**1.29. Solve: (a)** $(33.3)^2 = 1.11\times10^3$.

**(b)** $33.3\times45.1 = 1.50\times10^3$

Scientific notation is an easy way to establish significance.

**(c)** $\sqrt{22.2} - 1.2 = 3.5$

**(d)** $1/44.4 = 0.0225$

**1.35. Model:** Represent the watermelon as a particle for the motion diagram.

**Visualize:**

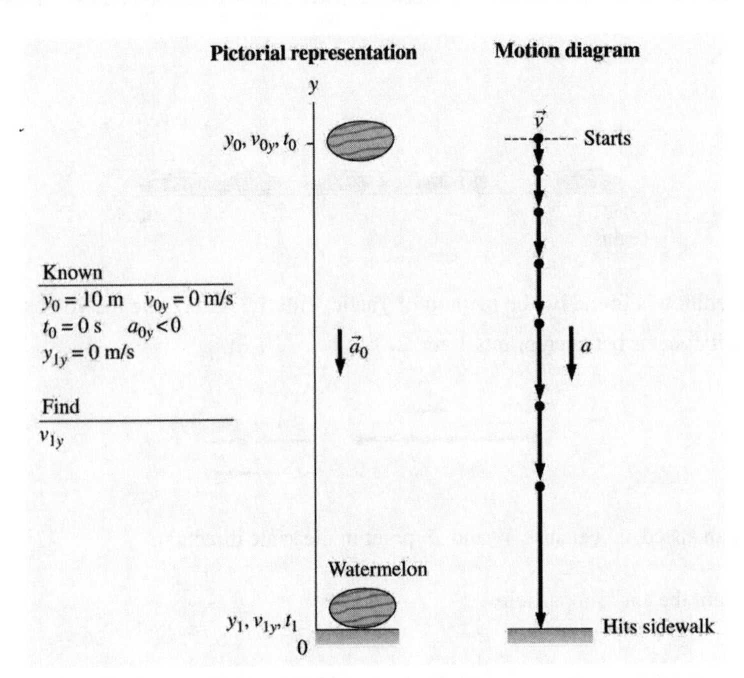


**1.39. Model:** Represent the ball as a particle for the motion diagram.

**Visualize:**

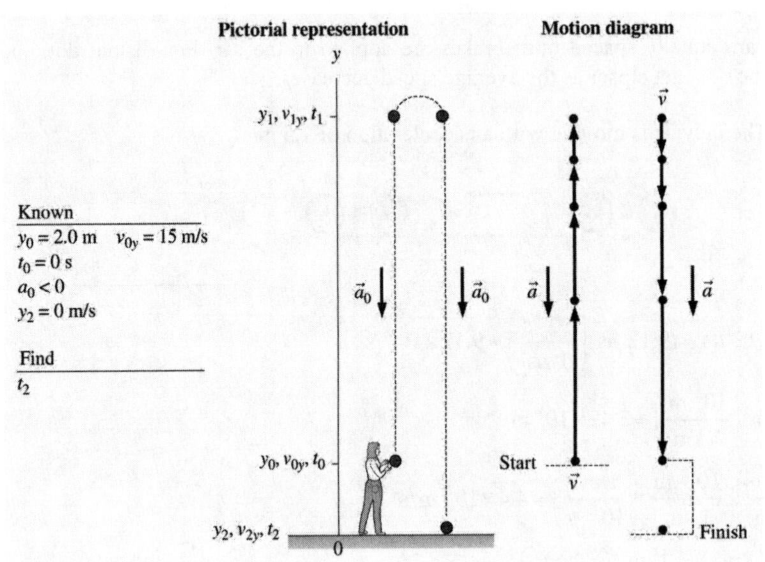

**1.49. Solve:**
**(a)**

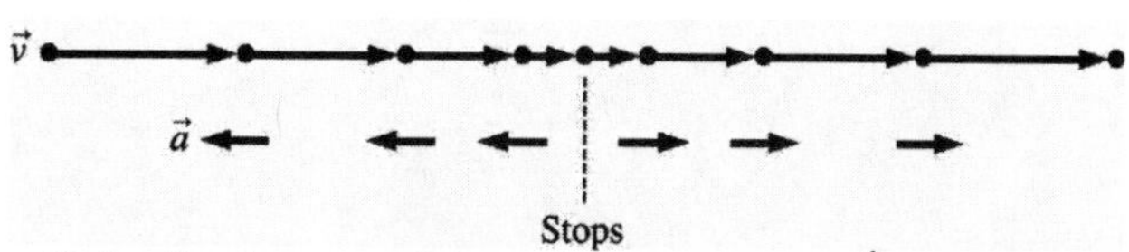


**(b)** Sue passes 3rd Street doing 30 km/h, slows steadily to the stop sign at 4th Street, stops for 1.0 s, then speeds up and reaches her original speed as she passes 5th Street. If the blocks are 50 m long, how long does it take Sue to drive from 3rd Street to 5th Street?
**(c)**

Known
$x_0 = 0$ m $t_0 = 0$ s
$v_{0x} = 20$ m/s
$a_{0x} = 0$ m/s$^2$
$v_{1x} = 2.0$ m/s
$t_1 = 0.50$ m/s
$v_{2x} = 0$ m/s $x_2 = 60$ m
Find
$a_{1x}$

**Pictorial representation**

Braking begins

Stops

$x_0, v_{0x}, t_0$ $\vec{a}_0 = 0$ $x_1, v_{1x}, t_1$ $\vec{a}_1$ $x_2, v_{2x}, t_2$ $x$

**1.51. Solve:**
**(a)**

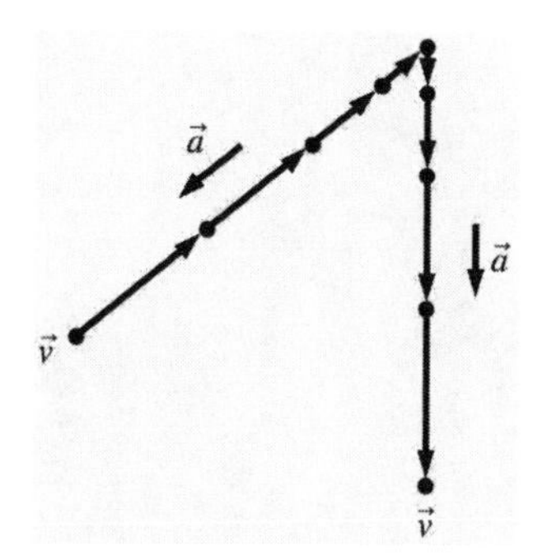

**(b)** Jeremy has perfected the art of steady acceleration and deceleration. From a speed of 60 mph he brakes his car to rest in 10 s with a constant deceleration. Then he turns into an adjoining street. Starting from rest, Jeremy accelerates with exactly the same magnitude as his earlier deceleration and reaches the same speed of 60 mph over the same distance in exactly the same time. Find the car's acceleration or deceleration.
**(c)**

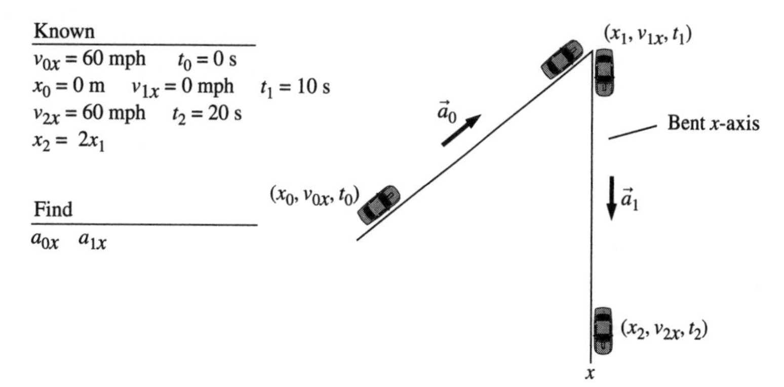

# 2

# KINEMATICS IN ONE DIMENSION

**2.1. Model:** We will consider the car to be a particle that occupies a single point in space.
**Visualize:**

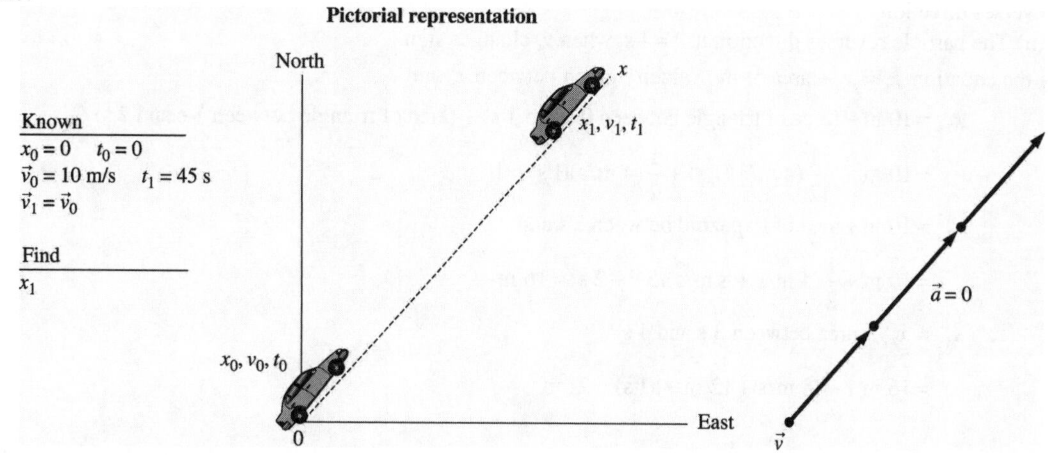


**Solve:** Since the velocity is constant, we have $x_f = x_i + v_x \Delta t$. Using the above values, we get

$$x_1 = 0 \text{ m} + (10 \text{ m/s})(45 \text{ s}) = 450 \text{ m}$$

**Assess:** 10 m/s ≈ 22 mph and implies a speed of 0.4 miles per minute. A displacement of 450 m in 45 s is reasonable and expected.

**2.3. Model:** Cars will be treated by the particle model.
**Visualize:**

Pictorial representation

Los Angeles — San Francisco

Known
$x_{\text{Alan }0} = x_{\text{Beth }0} = 0$
$x_{\text{Alan }1} = x_{\text{Beth }1} = 400$ miles
$v_{\text{Alan }0} = v_{\text{Alan }1} = 50$ mph
$v_{\text{Beth }0} = v_{\text{Beth }1} = 60$ mph
$t_{\text{Alan }0} = $ 8:00 AM
$t_{\text{Beth }0} = $ 9:00 AM

Find
$t_{\text{Alan }1}$ $t_{\text{Beth }1}$

$\vec{a} = 0$
$x_{\text{Alan }0}, v_{\text{Alan }0}, t_{\text{Alan }0}$
$x_{\text{Beth }0}, v_{\text{Beth }0}, t_{\text{Beth }0}$
$x_{\text{Alan }1}, v_{\text{Alan }1}, t_{\text{Alan }1}$
$x_{\text{Beth }1}, v_{\text{Beth }1}, t_{\text{Beth }1}$
x
8:00 Alan $\vec{v}$ $\vec{a} = 0$
9:00 Beth $\vec{v}$ $\vec{a} = 0$

**Solve:** Beth and Alan are moving at a constant speed, so we can calculate the time of arrival as follows:

$$v = \frac{\Delta x}{\Delta t} = \frac{x_1 - x_0}{t_1 - t_0} \Rightarrow t_1 = t_0 + \frac{x_1 - x_0}{v}$$

Using the known values identified in the pictorial representation, we find:

$$t_{\text{Alan 1}} = t_{\text{Alan 0}} + \frac{x_{\text{Alan 1}} - x_{\text{Alan 0}}}{v} = 8{:}00\text{ AM} + \frac{400\text{ mile}}{50\text{ miles/hour}} = 8{:}00\text{ AM} + 8\text{ hr} = 4{:}00\text{ PM}$$

$$t_{\text{Beth 1}} = t_{\text{Beth 0}} + \frac{x_{\text{Beth 1}} - x_{\text{Beth 0}}}{v} = 9{:}00\text{ AM} + \frac{400\text{ mile}}{60\text{ miles/hour}} = 9{:}00\text{ AM} + 6.67\text{ hr} = 3{:}40\text{ PM}$$

**(a)** Beth arrives first.
**(b)** Beth has to wait $t_{\text{Alan 1}} - t_{\text{Beth 1}} = 20$ minutesfor Alan.
**Assess:** Times of the order of 7 or 8 hours are reasonable in the present problem.

**2.7. Visualize:** Please refer to Figure EX2.7. The particle starts at $x_0 = 10$ m at $t_0 = 0$. Its velocity is initially in the $-x$ direction. The speed decreases as time increases during the first second, is zero at $t = 1$ s, and then increases after the particle reverses direction.
**Solve:** **(a)** The particle reverses direction at $t = 1$ s, when $v_x$ changes sign.
**(b)** Using the equation $x_f = x_0 +$ area of the velocity graph between $t_i$ and $t_f$,

$$x_{2\text{ s}} = 10\text{ m} - (\text{area of triangle between 0 s and 1 s}) + (\text{area of triangle between 1 s and 2 s})$$

$$= 10\text{ m} - \frac{1}{2}(4\text{ m/s})(1\text{ s}) + \frac{1}{2}(4\text{ m/s})(1\text{ s}) = 10\text{ m}$$

$$x_{3\text{ s}} = 10\text{ m} + \text{area of trapazoid between 2 s and 3 s}$$

$$= 10\text{ m} + \frac{1}{2}(4\text{ m/s} + 8\text{ m/s})(3\text{ s} - 2\text{ s}) = 16\text{ m}$$

$$x_{4\text{ s}} = x_{3\text{ s}} + \text{area between 3 s and 4 s}$$

$$= 16\text{ m} + \frac{1}{2}(8\text{ m/s} + 12\text{ m/s})(1\text{ s}) = 21\text{ m}$$

**2.13. Model:** We are using the particle model for the skater and the kinematics model of motion under constant acceleration.
**Solve:** Since we don't know the time of acceleration we will use

$$v_f^2 = v_i^2 + 2a(x_f - x_i)$$

$$\Rightarrow a = \frac{v_f^2 - v_i^2}{2(x_f - x_i)} = \frac{(6.0\text{ m/s})^2 - (8.0\text{ m/s})^2}{2(5.0\text{ m})} = -2.8\text{ m/s}^2$$

**Assess:** A deceleration of 2.8 m/s$^2$ is reasonable.

**2.15. Model:** Represent the spherical drop of molten metal as a particle.
**Visualize:**

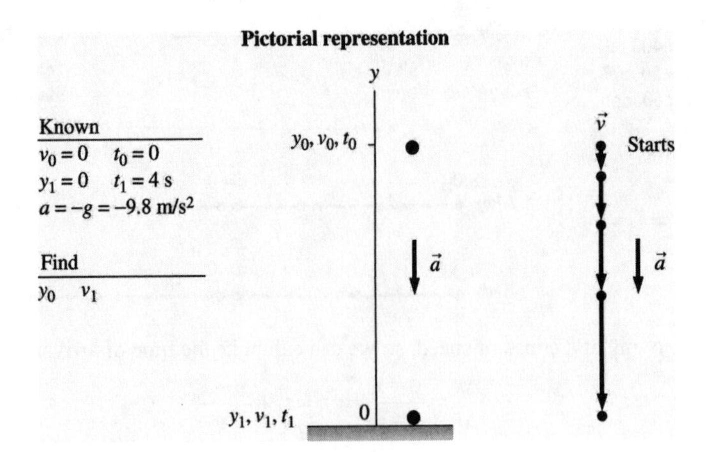

**Solve:** **(a)** The shot is in free fall, so we can use free fall kinematics with $a=-g$. The height must be such that the shot takes 4 s to fall, so we choose $t_1 = 4$ s. Then,

$$y_1 = y_0 + v_0(t_1 - t_0) - \frac{1}{2}g(t_1 - t_0)^2 \Rightarrow y_0 = \frac{1}{2}gt_1^2 = \frac{1}{2}(9.8 \text{ m/s}^2)(4 \text{ s})^2 = 78.4 \text{ m}$$

**(b)** The impact velocity is $v_1 = v_0 - g(t_1 - t_0) = -gt_1 = -39.2$ m/s.

**Assess:** Note the minus sign. The question asked for *velocity*, not speed, and the *y*-component of $\vec{v}$ is negative because the vector points downward.

**2.19. Model:** We will represent the skier as a particle.

**Visualize:**

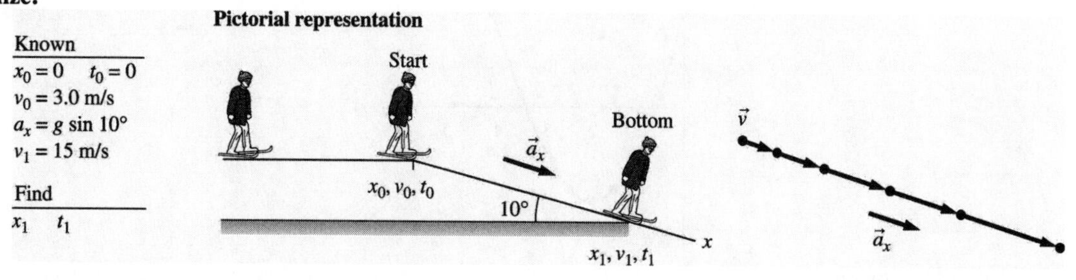


Note that the skier's motion on the horizontal, frictionless snow is not of any interest to us. Also note that the acceleration parallel to the incline is equal to $g \sin 10°$.

**Solve:** Using the following constant-acceleration kinematic equations,

$$v_{fx}^2 = v_{ix}^2 + 2a_x(x_f - x_i)$$

$$\Rightarrow (15 \text{ m/s})^2 = (3.0 \text{ m/s})^2 + 2(9.8 \text{ m/s}^2)\sin 10°(x_1 - 0 \text{ m}) \Rightarrow x_1 = 64 \text{ m}$$

$$v_{fx} = v_{ix} + a_x(t_f - t_i)$$

$$\Rightarrow (15 \text{ m/s}) = (3.0 \text{ m/s}) + (9.8 \text{ m/s}^2)(\sin 10°)t \Rightarrow t = 7.1 \text{ s}$$

**Assess:** A time of 7.1 s to cover 64 m is a reasonable value.

**2.23. Solve:** The formula for the particle's velocity is given by

$$v_f = v_i + \text{area under the acceleration curve between } t_i \text{ and } t_f$$

For $t = 4$ s, we get

$$v_{4\text{ s}} = 8 \text{ m/s} + \frac{1}{2}(4 \text{ m/s}^2)4 \text{ s} = 16 \text{ m/s}$$

**Assess:** The acceleration is positive but decreases as a function of time. The initial velocity of 8.0 m/s will therefore increase. A value of 16 m/s is reasonable.

**2.25. Solve:** **(a)**

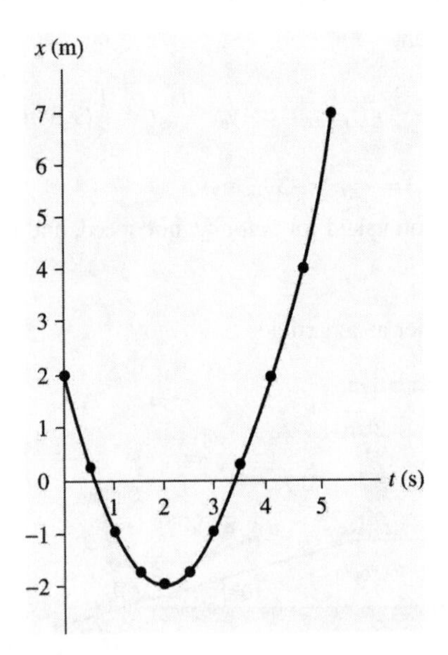


**(b)** To be completed by student.

**(c)** $\dfrac{dx}{dt} = v_x = 2t - 4 \Rightarrow v_x(\text{at } t = 1\text{ s}) = [2\text{ m/s}^2(1\text{ s}) - 4\text{ m/s}] = -2\text{ m/s}$

**(d)** There is a turning point at $t = 2$ s.

**(e)** Using the equation in part (c),

$$v_x = 4\text{ m/s} = (2t - 4)\text{ m/s} \Rightarrow t = 4$$

Since $x = (t^2 - 4t + 2)$ m, $x = 2$ m.

**(f)**

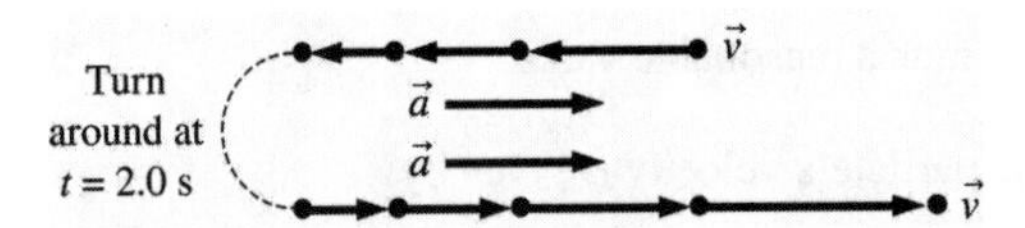


**2.27. Visualize:** Please refer to Figure P2.27.

**Solve:** **(a)** We can calculate the position of the particle at every instant with the equation

$$x_f = x_i + \text{area under the velocity-versus-time graph between } t_i \text{ and } t_f$$

The particle starts from the origin at $t = 0$ s, so $x_i = 0$ m. Notice that the each square of the grid in Figure P2.27 has "area" $(5\text{ m/s}) \times (2\text{ s}) = 10$ m. We can find the area under the curve, and thus determine $x$, by counting squares. You can see that $x = 35$ m at $t = 4$ s because there are 3.5 squares under the curve. In addition, $x = 35$ m at $t = 8$ s because the 5 m represented by the half square between 4 and 6 s is cancelled by the –5 m represented by the half square between 6 and 8 s. Areas beneath the axis are negative areas. The particle passes through $x = 35$ m at $t = 4$ s and again at $t = 8$ s.

**(b)** The particle moves to the right for $0\text{ s} \geq t \geq 6$ s, where the velocity is positive. It reaches a turning point at $x = 40$ m at $t = 6$ s. The motion is to the left for $t > 6$ s. This is shown in the motion diagram below.

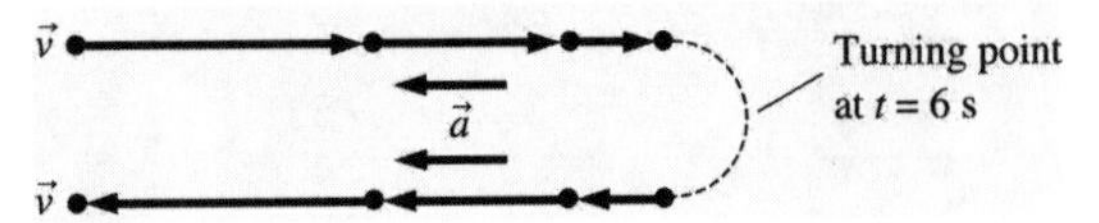


**2.35. Model:** Represent the ball as a particle.

**Visualize:** Please refer to Figure P2.35.

**Solve:** In the first and third segments the acceleration $a_s$ is zero. In the second segment the acceleration is negative and constant. This means the velocity $v_s$ will be constant in the first two segments and will decrease linearly in the thirdsegment. Because the velocity is constant in the first and third segments, the position $s$ will increase linearly. In the second segment, the position will increase parabolically rather than linearly because the velocity decreases linearly with time.

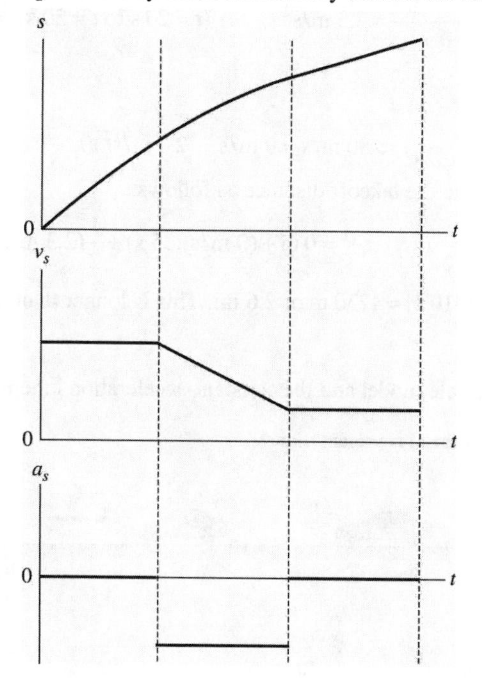


**2.37. Model:** Represent the ball as a particle.
**Visualize:** Please refer to Figure P2.37. The ball moves to the right along the first track until it strikes the wall, which causes it to move to the left on a second track. The ball then descends on a third track until it reaches the fourth track, which is horizontal.
**Solve:**

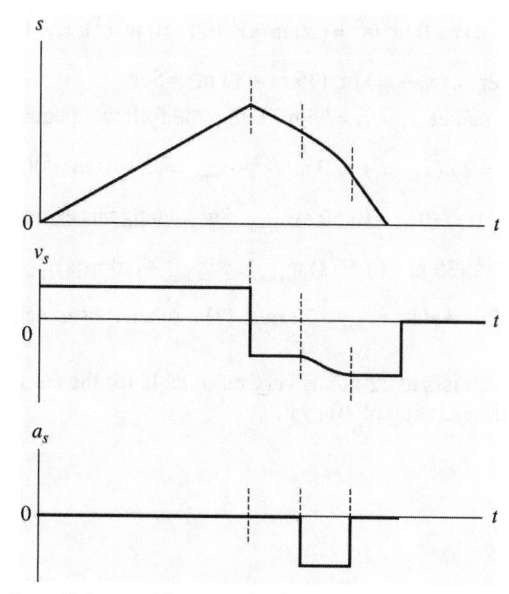


**Assess:** Note that the time derivative of the position graph yields the velocity graph, and the derivative of the velocity graph gives the acceleration graph.

**2.41. Model:** The plane is a particle and the constant-acceleration kinematic equations hold.
**Solve:** **(a)** To convert 80 m/s to mph, we calculate 80 m/s × 1 mi/1609 m × 3600 s/h = 179 mph.

**(b)** Using $a_s = \Delta v/\Delta t$ , we have,

$$a_s(t=0 \text{ to } t=10\text{ s}) = \frac{23\text{ m/s}-0\text{ m/s}}{10\text{ s}-0\text{ s}} = 2.3\text{ m/s}^2 \qquad a_s(t=20\text{ s to } t=30\text{ s}) = \frac{69\text{ m/s}-46\text{ m/s}}{30\text{ s}-20\text{ s}} = 2.3\text{ m/s}^2$$

For all time intervals $a$ is 2.3 m/s$^2$.
**(c)** Using kinematics as follows:

$$v_{\text{fs}} = v_{\text{is}} + a(t_{\text{f}} - t_{\text{i}}) \Rightarrow 80\text{ m/s} = 0\text{ m/s} + (2.3\text{ m/s}^2)(t_{\text{f}} - 0\text{ s}) \Rightarrow t_{\text{f}} = 35\text{ s}$$

**(d)** Using the above values, we calculate the takeoff distance as follows:

$$s_{\text{f}} = s_{\text{i}} + v_{\text{is}}(t_{\text{f}} - t_{\text{i}}) + \frac{1}{2}a_{\text{s}}(t_{\text{f}} - t_{\text{i}})^2 = 0\text{ m} + (0\text{ m/s})(35\text{ s}) + \frac{1}{2}(2.3\text{ m/s}^2)(35\text{ s})^2 = 1410\text{ m}$$

For safety, the runway should be $3\times1410\text{ m} = 4230\text{ m}$ or 2.6 mi. This is longer than the 2.4 mi long runway, so the takeoff is not safe.

**2.47. Model:** We will use the particle model and the constant-acceleration kinematic equations.
**Visualize:**

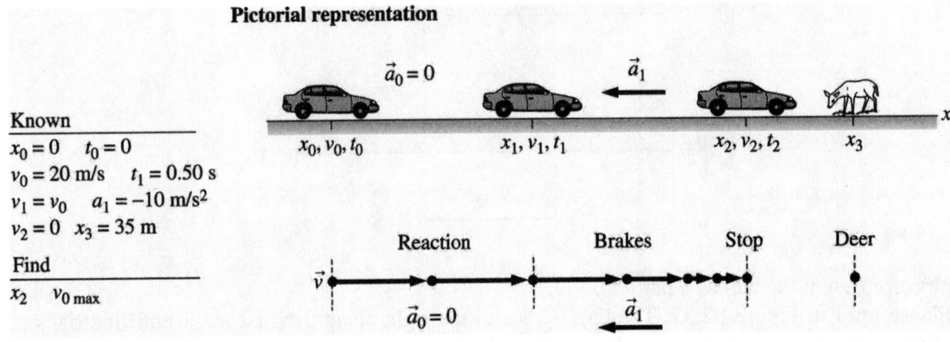


**Solve:** **(a)** To find $x_2$, we first need to determine $x_1$. Using $x_1 = x_0 + v_0(t_1 - t_0)$, we get $x_1 = 0\text{ m} + (20\text{ m/s})(0.50\text{ s} - 0\text{ s}) =$ 10 m. Now,

$$v_2^2 = v_1^2 + 2a_1(x_2 - x_1) \Rightarrow 0\text{ m}^2/\text{s}^2 = (20\text{ m/s})^2 + 2(-10\text{ m/s}^2)(x_2 - 10\text{ m}) \Rightarrow x_2 = 30\text{ m}$$

The distance between you and the deer is $(x_3 - x_2)$ or $(35\text{ m} - 30\text{ m}) = 5\text{ m}$.
**(b)** Let us find $v_{0\,\max}$ such that $v_2 = 0$ m/s at $x_2 = x_3 = 35$ m. Using the following equation,

$$v_2^2 - v_{0\,\max}^2 = 2a_1(x_2 - x_1) \Rightarrow 0\text{ m}^2/\text{s}^2 - v_{0\,\max}^2 = 2(-10\text{ m/s}^2)(35\text{ m} - x_1)$$

Also, $x_1 = x_0 + v_{0\,\max}(t_1 - t_0) = v_{0\,\max}(0.50\text{ s} - 0\text{ s}) = (0.50\text{ s})v_{0\,\max}$. Substituting this expression for $x_1$ in the above equation yields

$$-v_{0\,\max}^2 = (-20\text{ m/s}^2)[35\text{ m} - (0.50\text{ s})\, v_{0\,\max}] \Rightarrow v_{0\,\max}^2 + (10\text{ m/s})v_{0\,\max} - 700\text{ m}^2/\text{s}^2 = 0$$

The solution of this quadratic equation yields $v_{0\,\max} = 22$ m/s. (The other root is negative and unphysical for the present situation.)
**Assess:** An increase of speed from 20 m/s to 22 m/s is very reasonable for the car to cover an additional distance of 5 m with a reaction time of 0.50 s and a deceleration of 10 m/s$^2$.

**2.53. Model:** The car is a particle moving under constant-acceleration kinematic equations.
**Visualize:**

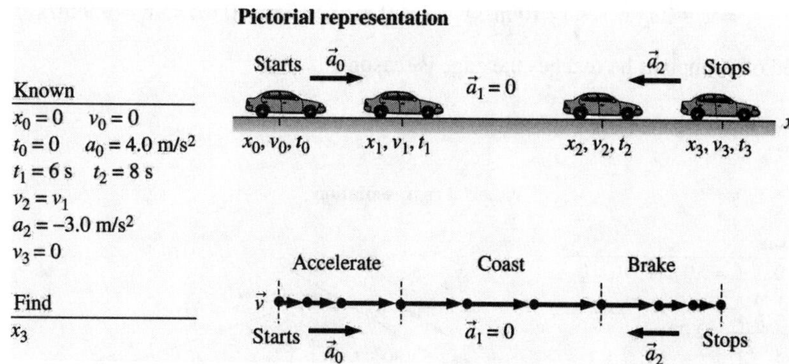


**Solve:** This is a three-part problem. First the car accelerates, then it moves with a constant speed, and then it decelerates. First, the car accelerates:

$$v_1 = v_0 + a_0(t_1 - t_0) = 0 \text{ m/s} + (4.0 \text{ m/s}^2)(6 \text{ s} - 0 \text{ s}) = 24 \text{ m/s}$$

$$x_1 = x_0 + v_0(t_1 - t_0) + \frac{1}{2}a_0(t_1 - t_0)^2 = 0 \text{ m} + \frac{1}{2}(4.0 \text{ m/s}^2)(6 \text{ s} - 0 \text{ s})^2 = 72 \text{ m}$$

Second, the car moves at $v_1$:

$$x_2 - x_1 = v_1(t_2 - t_1) + \frac{1}{2}a_1(t_2 - t_1)^2 = (24 \text{ m/s})(8 \text{ s} - 6 \text{ s}) + 0 \text{ m} = 48 \text{ m}$$

Third, the car decelerates:

$$v_3 = v_2 + a_2(t_3 - t_2) \Rightarrow 0 \text{ m/s} = 24 \text{ m/s} + (-3.0 \text{ m/s}^2)(t_3 - t_2) \Rightarrow (t_3 - t_2) = 8 \text{ s}$$

$$x_3 = x_2 + v_2(t_3 - t_2) + \frac{1}{2}a_2(t_3 - t_2)^2 \Rightarrow x_3 - x_2 = (24 \text{ m/s})(8 \text{ s}) + \frac{1}{2}(-3.0 \text{ m/s}^2)(8 \text{ s})^2 = 96 \text{ m}$$

Thus, the total distance between stop signs is:

$$x_3 - x_0 = (x_3 - x_2) + (x_2 - x_1) + (x_1 - x_0) = 96 \text{ m} + 48\text{m} + 72 \text{ m} = 216 \text{ m}$$

**Assess:** A distance of approximately 600 ft in a time of around 10 s with an acceleration/deceleration of the order of 7 mph/s is reasonable.

**2.55. Model:** Santa is a particle moving under constant-acceleration kinematic equations.
**Visualize:** Note that our $x$-axis is positioned along the incline.

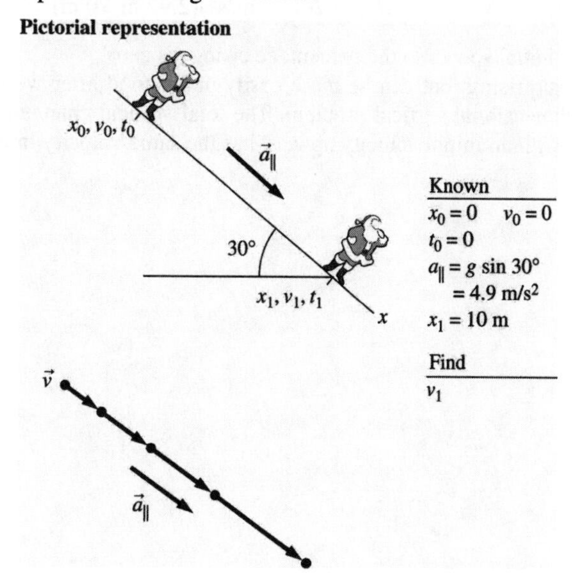

**Solve:** Using the following kinematic equation,

$$v_1^2 = v_0^2 + 2a_{\parallel}(x_1 - x_2) = (0 \text{ m/s})^2 + 2(4.9 \text{ m/s}^2)(10 \text{ m} - 0 \text{ m}) \Rightarrow v_1 = 9.9 \text{ m/s}$$

**Assess:** Santa's speed of 20 mph as he reaches the edge is reasonable.

**2.57. Model:** We will use the particle model for the puck.
**Visualize:**

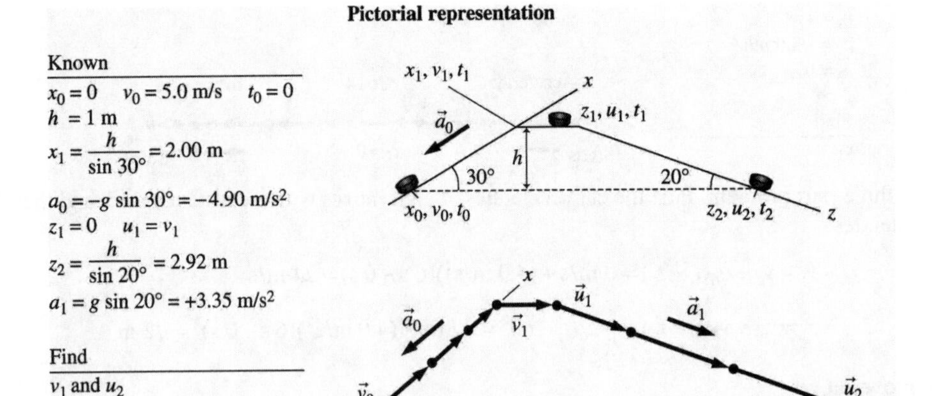


We can view this problem as two one-dimensional motion problems. The horizontal segments do not affect the motion because the speed does not change. So, the problem "starts" at the bottom of the uphill ramp and "ends" at the bottom of the downhill ramp. At the top of the ramp the speed does not change along the horizontal section. The final speed from the uphill roll (first problem) becomes the initial speed of the downhill roll (second problem). Because the axes point in different directions, we can avoid possible confusion by calling the downhill axis the $z$-axis and the downhill velocities $u$. The uphill axis as usual will be denoted by $x$ and the uphill velocities as $v$. Note that the height information, $h = 1$ m, has to be transformed into information about positions along the two axes.

**Solve:** **(a)** The uphill roll has $a_0 = -g\sin 30° = -4.90 \text{ m/s}^2$. The speed at the top is found from

$$v_1^2 = v_0^2 + 2a_0(x_1 - x_0)$$

$$\Rightarrow v_1 = \sqrt{v_0^2 + 2a_0 x_1} = \sqrt{(5 \text{ m/s})^2 + 2(-4.90 \text{ m/s}^2)(2.00 \text{ m})} = 2.32 \text{ m/s}$$

**(b)** The downward roll starts with velocity $u_1 = v_1 = 2.32$ m/s and $a_1 = +g\sin 20° = 3.35 \text{ m/s}^2$. Then,

$$u_2^2 = u_1^2 + 2a_1(z_2 - z_1) = (2.32 \text{ m/s})^2 + 2(3.35 \text{ m/s}^2)(2.92 \text{ m} - 0 \text{ m}) \Rightarrow u_2 = 5.00 \text{ m/s}$$

**(c)** The final speed is equal to the initial speed, so the percentage change is zero!

**Assess:** This result may seem surprising, but can be more easily understood after we introduce the concept of energy. For now, imagine this is a one dimensional vertical problem. The total vertical change in height of the puck is zero. We have already seen how an object with an initial velocity upward has the same velocity in the opposite direction as it passes through that height going down.

**2.63. Model:** The ball is a particle that exhibits freely falling motion according to the constant-acceleration kinematic equations.
**Visualize:**

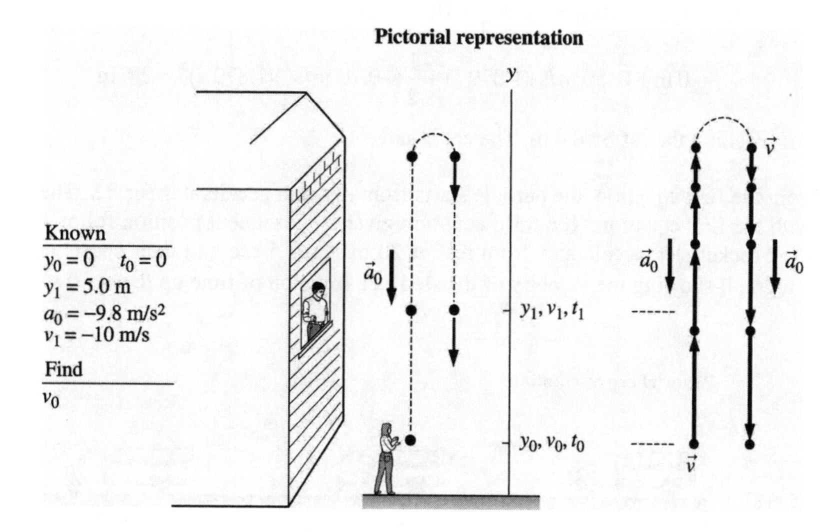


**Solve:** Using the known values, we have

$$v_1^2 = v_0^2 + 2a_0(y_1 - y_0) \Rightarrow (-10 \text{ m/s})^2 = v_0^2 + 2(-9.8 \text{ m/s}^2)(5.0 \text{ m} - 0 \text{ m}) \Rightarrow v_0 = 14 \text{ m/s}$$

**2.67. Model:** We will represent the dog and the cat in the particle model.
**Visualize:**

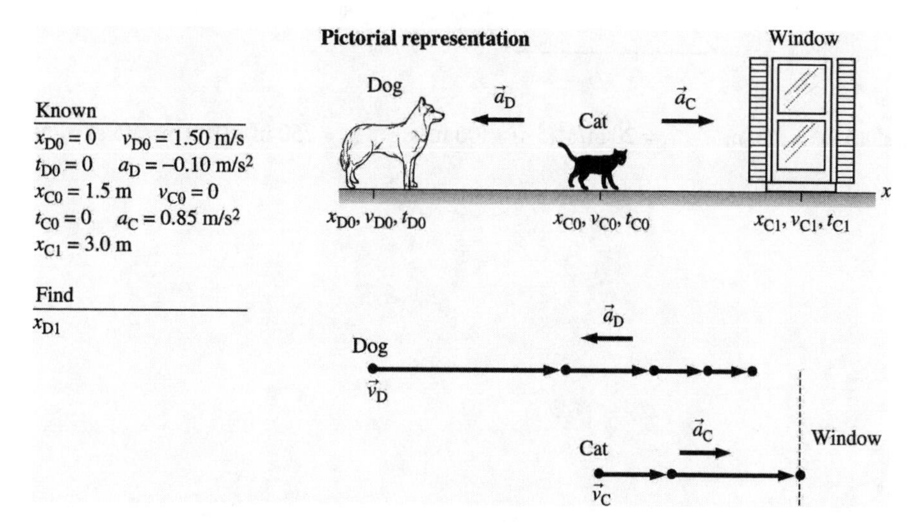


**Solve:** We will first calculate the time $t_{C1}$ the cat takes to reach the window. The dog has exactly the same time to reach the cat (or the window). Let us therefore first calculate $t_{C1}$ as follows:

$$x_{C1} = x_{C0} + v_{C0}(t_{C1} - t_{C0}) + \frac{1}{2}a_C(t_{C1} - t_{C0})^2$$

$$\Rightarrow 3.0 \text{ m} = 1.5 \text{ m} + 0 \text{ m} + \frac{1}{2}(0.85 \text{ m/s}^2)t_{C1}^2 \Rightarrow t_{C1} = 1.879 \text{ s}$$

In the time $t_{D1} = 1.879$ s, the dog's position can be found as follows:

$$x_{D1} = x_{D0} + v_{D0}(t_{D1} - t_{D0}) + \frac{1}{2}a_D(t_{D1} - t_{D0})^2$$
$$= 0 \text{ m} + (1.50 \text{ m/s})(1.879 \text{ s}) + \frac{1}{2}(-0.10 \text{ m/s}^2)(1.879 \text{ s})^2 = 2.6 \text{ m}$$

That is, the dog is shy of reaching the cat by 0.4 m. The cat is safe.

**2.75. Solve:** **(a)** From the first equation, the particle starts from rest and accelerates for 5 s. The second equation gives a position consistent with the first equation. The third equation gives a subsequent position following the second equation with zero acceleration. A rocket sled accelerates from rest at 20 m/s$^2$ for 5 sec and then coasts at constant speed for an additional 5 sec. Draw a graph showing the velocity of the sled as a function of time up to $t = 10$ s. Also, how far does the sled move in 10 s?

**(b)**

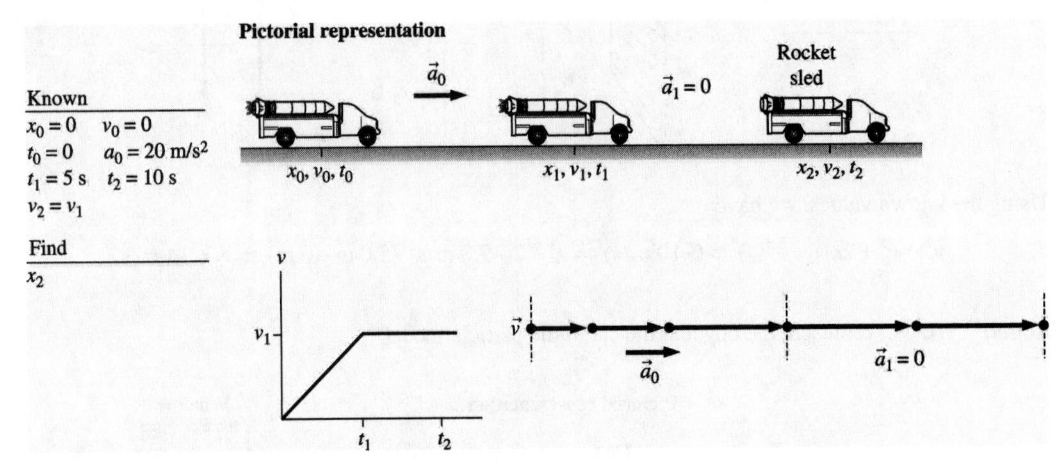


**(c)** $x_1 = \frac{1}{2}(20 \text{ m/s}^2)(5 \text{ s})^2 = 250 \text{ m}$ $\quad v_1 = 20 \text{ m/s}^2(5 \text{ s}) = 100 \text{ m/s}$ $\quad x_2 = 250 \text{ m} + (100 \text{ m/s})(5 \text{ s}) = 750 \text{ m}$

# VECTORS AND COORDINATE SYSTEMS

3

**3.5. Visualize:** The figure shows the components $v_x$ and $v_y$, and the angle $\theta$.

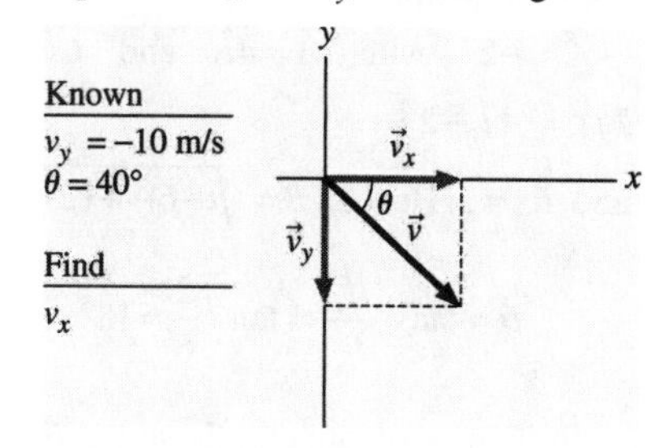


**Solve:** We have, $v_y = -v\sin 40°$, or $-10 \text{ m/s} = -v\sin 40°$, or $v = 15.56$ m/s.

Thus the $x$-component is $v_x = v\cos 40° = (15.56 \text{ m/s}) \cos 40° = 12$ m/s.

**Assess:** The $x$-component is positive since the position vector is in the fourth quadrant.

**3.11. Visualize:**

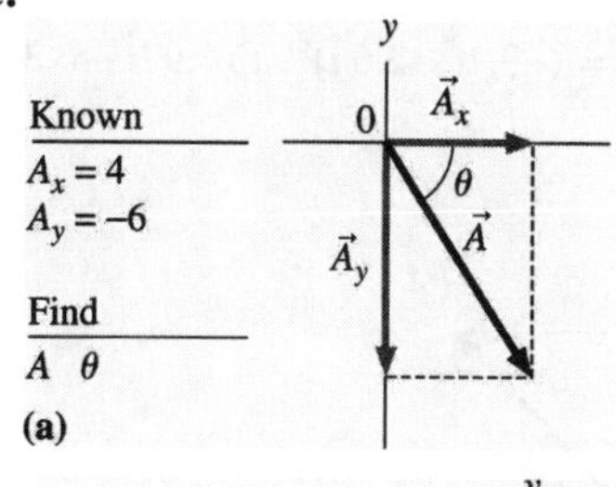


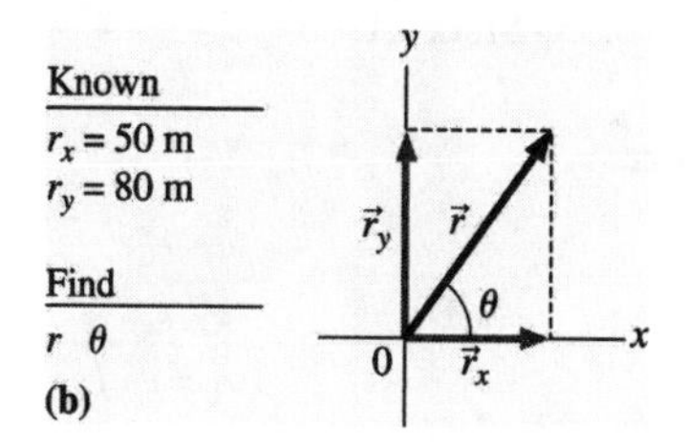


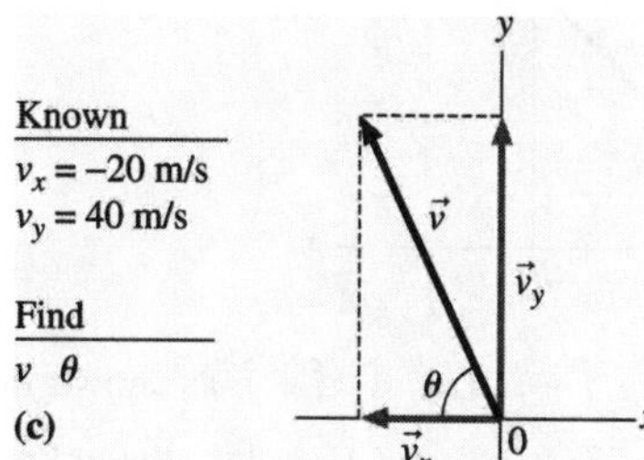


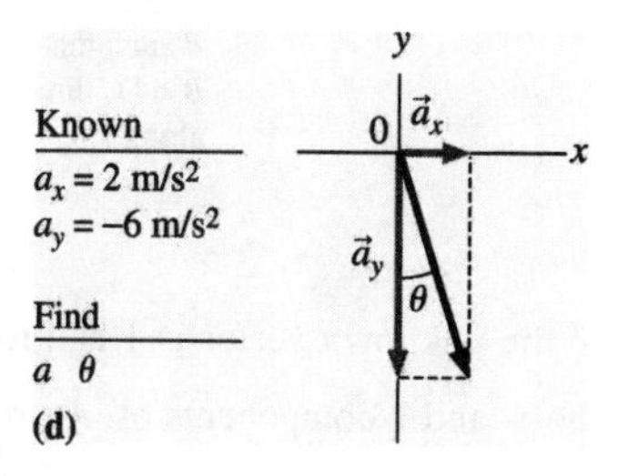


**Solve:** **(a)** Using the formulas for the magnitude and direction of a vector, we have:

$$A = \sqrt{(4)^2 + (-6)^2} = 7.21 \qquad \theta = \tan^{-1}\frac{6}{4} = \tan^{-1}1.5 = 56.3°$$

**(b)** $r = \sqrt{(50 \text{ m})^2 + (80 \text{ m})^2} = 94.3 \text{ m} \qquad \theta = \tan^{-1}\left(\frac{r_y}{r_x}\right) = \tan^{-1}\left(\frac{80 \text{ m}}{50 \text{ m}}\right) = 58.0°$

**(c)** $v = \sqrt{(-20 \text{ m/s})^2 + (40 \text{ m/s})^2} = 44.7 \text{ m/s} \qquad \theta = \tan^{-1}\frac{40}{20} = \tan^{-1}2 = 63.4°$

**(d)** $a = \sqrt{(2\text{ m/s}^2)^2 + (-6\text{ m/s}^2)^2} = 6.3\text{ m/s}^2$ $\qquad \theta = \tan^{-1}\dfrac{2}{6} = \tan^{-1}0.33 = 18.4°$

**Assess:** Note that the angle $\theta$ made with the $x$-axis is smaller than 45° whenever $|E_y| < |E_x|$, $\theta = 45°$ for $|E_y| = |E_x|$, and $\theta > 45°$ for $|E_y| > |E_x|$. In part (d), $\theta$ is with the $y$-axis, where the opposite of this rule applies.

**3.19. Visualize:** Refer to Figure EX3.19. The velocity vector $\vec{v}$ points west and makes an angle of 30° with the $-x$-axis. $\vec{v}$ points to the left and up, implying that $v_x$ is negative and $v_y$ is positive.

**Solve:** We have $v_x = -v\cos 30° = -(100\text{ m/s})\cos 30° = -86.6\text{ m/s}$ and $v_y = +v\sin 30° = (100\text{ m/s})\sin 30° = 50.0\text{ m/s}$.

**Assess:** $v_x$ and $v_y$ have the same units as $\vec{v}$.

**3.25. Visualize:** Refer to Figure P3.25 in your textbook.

**Solve:** **(a)** We are given that $\vec{A} + \vec{B} + \vec{C} = -2\hat{\imath}$ with $\vec{A} = 4\hat{\imath}$, and $\vec{C} = -2\hat{\jmath}$. This means $\vec{A} + \vec{C} = 4\hat{\imath} - 2\hat{\jmath}$. Thus, $\vec{B} = (\vec{A} + \vec{B} + \vec{C}) - (\vec{A} + \vec{C}) = (-2\hat{\imath}) - (4\hat{\imath} - 2\hat{\jmath}) = -6\hat{\imath} + 2\hat{\jmath}$.

**(b)** We have $\vec{B} = B_x\hat{\imath} + B_y\hat{\jmath}$ with $B_x = -6$ and $B_y = 2$. Hence, $B = \sqrt{(-6)^2 + (2)^2} = 6.3$

$$\theta = \tan^{-1}\frac{B_y}{|B_x|} = \tan^{-1}\frac{2}{6} = 18°$$

Since $\vec{B}$ has a negative $x$-component and a positive $y$-component, the angle $\theta$ made by $\vec{B}$ is with the $-x$-axis and it is above the $-x$-axis.

**Assess:** Since $|B_y| < |B_x|$, $\theta < 45°$ as is obtained above.

**3.27. Visualize:** Refer to Figure P3.27.

**Solve:** From the rules of trigonometry, we have $A_x = 4\cos 40° = 3.1$ and $A_y = 4\sin 40° = 2.6$. Also, $B_x = -2\cos 10° = -1.97$ and $B_y = +2\sin 10° = 0.35$. Since $\vec{A} + \vec{B} + \vec{C} = \vec{0}$, $\vec{C} = -\vec{A} - \vec{B} = (-\vec{A}) + (-\vec{B}) = (-3.1\hat{\imath} - 2.6\hat{\jmath}) + (+1.97\hat{\imath} - 0.35\hat{\jmath}) = -1.1\hat{\imath} - 3.0\hat{\jmath}$.

**3.29. Visualize:**

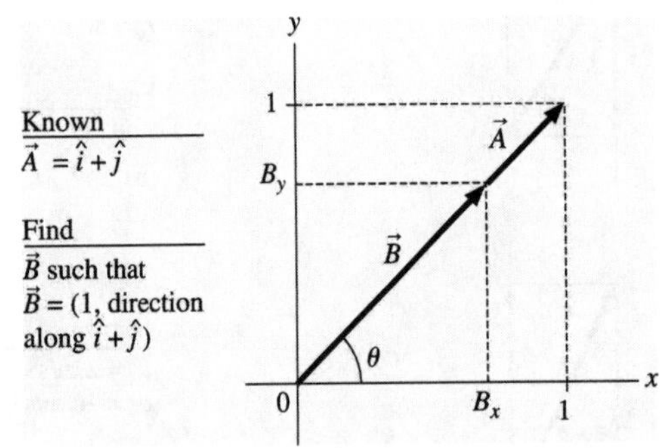


The magnitude of the unknown vector is 1 and its direction is along $\hat{\imath} + \hat{\jmath}$. Let $\vec{A} = \hat{\imath} + \hat{\jmath}$ as shown in the diagram. That is, $\vec{A} = 1\hat{\imath} + 1\hat{\jmath}$ and the $x$- and $y$-components of $\vec{A}$ are both unity. Since $\theta = \tan^{-1}(A_y/A_x) = 45°$, the unknown vector must make an angle of 45° with the $+x$-axis and have unit magnitude.

**Solve:** Let the unknown vector be $\vec{B} = B_x\hat{\imath} + B_y\hat{\jmath}$ where

$$B_x = B\cos 45° = \frac{1}{\sqrt{2}}B \quad \text{and} \quad B_y = B\sin 45° = \frac{1}{\sqrt{2}}B$$

We want the magnitude of $\vec{B}$ to be 1, so we have

$$B = \sqrt{B_x^2 + B_y^2} = 1$$

$$\Rightarrow \sqrt{\left(\frac{1}{\sqrt{2}}B\right)^2 + \left(\frac{1}{\sqrt{2}}B\right)^2} = 1 \Rightarrow \sqrt{B^2} = 1 \Rightarrow B = 1$$

Hence,

$$B_x = B_y = \frac{1}{\sqrt{2}}$$

Finally,

$$\vec{B} = B_x\hat{i} + B_y\hat{j} = \frac{1}{\sqrt{2}}\hat{i} + \frac{1}{\sqrt{2}}\hat{j}$$

**3.33. Visualize:** **(a)**

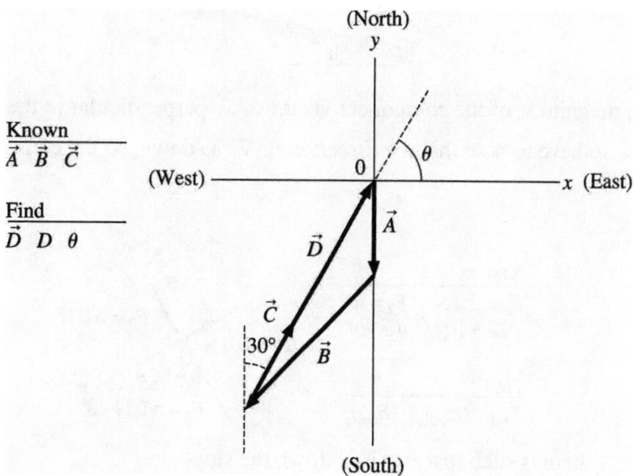


Note that $+x$ is along the east and $+y$ is along the north.

**Solve:** **(b)** We are given $\vec{A} = -(200\text{ m})\hat{j}$, and can use trigonometry to obtain $\vec{B} = -(283\text{ m})\hat{i} - (283\text{ m})\hat{j}$ and $\vec{C} = (100\text{ m})\hat{i} + (173\text{ m})\hat{j}$. We want $\vec{A} + \vec{B} + \vec{C} + \vec{D} = 0$. This means

$$\vec{D} = -\vec{A} - \vec{B} - \vec{C}$$
$$= (200\text{ m}\hat{j}) + (283\text{ m}\,\hat{i} + 283\text{ m}\hat{j}) + (-100\text{ m}\,\hat{i} - 173\text{ m}\hat{j}) = 183\text{ m}\,\hat{i} + 310\text{ m}\hat{j}$$

The magnitude and direction of $\vec{D}$ are

$$D = \sqrt{(183\text{ m})^2 + (310\text{ m})^2} = 360\text{ m} \quad \text{and} \quad \theta = \tan^{-1}\frac{D_y}{D_x} = \tan^{-1}\left(\frac{310\text{ m}}{183\text{ m}}\right) = 59.4°$$

This means $\vec{D} = (360\text{ m}, 59.4°$ north of east).

**(c)** The measured length of the vector $\vec{D}$ on the graph (with a ruler) is approximately 1.75 times the measured length of vector $\vec{A}$. Since $A = 200$ m, this gives $D = 1.75 \times 200\text{ m} = 350$ m. Similarly, the angle $\theta$ measured with the protractor is close to 60°. These answers are in close agreement to part (b).

**3.37. Visualize:**

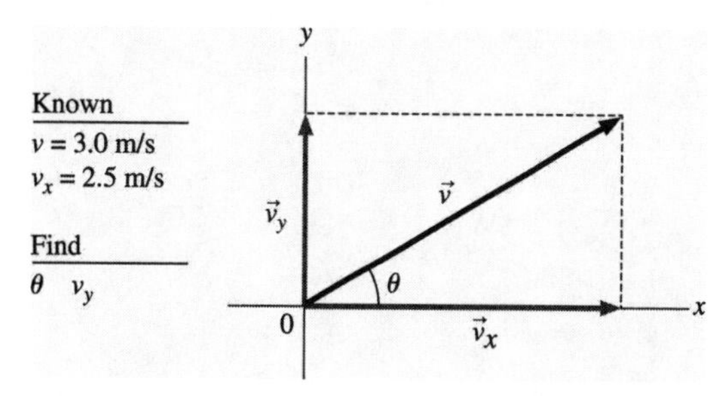


**Solve:** **(a)** Since $v_x = v\cos\theta$, we have $2.5\text{ m/s} = (3.0\text{ m/s})\cos\theta \Rightarrow \theta = \cos^{-1}\left(\frac{2.5\text{ m/s}}{3.0\text{ m/s}}\right) = 34°$.

**(b)** The vertical component is $v_y = v\sin\theta = (3.0\text{ m/s})\sin 34° = 1.7\text{ m/s}$.

**3.41. Visualize:** A 3% grade rises 3 m for every 100 m horizontal distance. The angle of the ground is thus $\alpha = \tan^{-1}(3/100) = \tan^{-1}(0.03) = 1.72°$.
Establish a tilted coordinate system with one axis parallel to the ground and the other axis perpendicular to the ground.

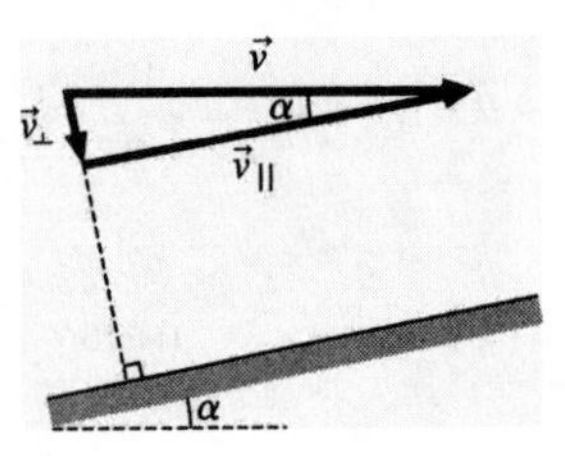


**Solve:** From the figure, the magnitude of the component vector of $\vec{v}$ perpendicular to the ground is $v_\perp = v\sin\alpha = 15.0$ m/s. But this is only the size. We also have to note that the *direction* of $\vec{v}_\perp$ is down, so the component is $v_\perp = -15.0$ m/s.

**3.45. Visualize:**

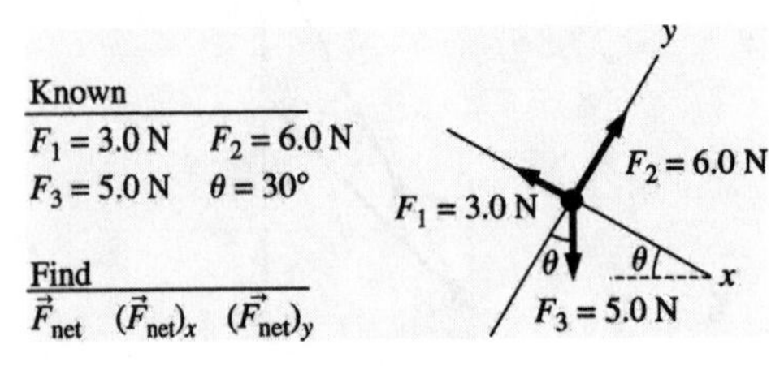


Note that the tilted coordinate system is such that $x$-axis is down the slope.
**Solve:** Expressing all three forces in terms of unit vectors, we have $\vec{F}_1 = -(3.0\text{ N})\hat{i}$, $\vec{F}_2 = +(6.0\text{ N})\hat{j}$, and $\vec{F}_3 = (5.0\text{ N})\sin\theta\hat{i} - (5.0\text{ N})\cos\theta\hat{j}$.
**(a)** The component of $\vec{F}_{net}$ parallel to the floor is $(\vec{F}_{net})_x = (\vec{F}_1 + \vec{F}_2 + \vec{F}_3)_x = -(3.0\text{ N})\hat{i} + (5.0\text{ N})\sin 30°\hat{i} = -(0.50\text{ N})\hat{i}$.
**(b)** The component of $\vec{F}_{net}$ perpendicular to the floor is $(\vec{F}_{net})_y = (\vec{F}_1 + \vec{F}_2 + \vec{F}_3)_y = (6.0\text{ N})\hat{j} - (5.0\text{ N})\cos 30°\hat{j} = (1.67\text{ N})\hat{j}$.
**(c)** The magnitude of $\vec{F}_{net}$ is $F_{net} = \sqrt{(F_{net})_x + (F_{net})_y} = \sqrt{(-0.50\text{ N})^2 + (1.67\text{ N})^2} = 1.74$ N. The angle $\vec{F}_{net}$ makes is

$$\theta = \tan^{-1}\frac{(F_{net})_y}{|(\vec{F}_{net})_x|} = \tan^{-1}\left(\frac{1.67\text{ N}}{0.50\text{ N}}\right) = 73°$$

$\vec{F}_{net}$ is 73° above the $-x$-axis in quadrant II.

# KINEMATICS IN TWO DIMENSIONS

4

**4.3. Solve: (a)**

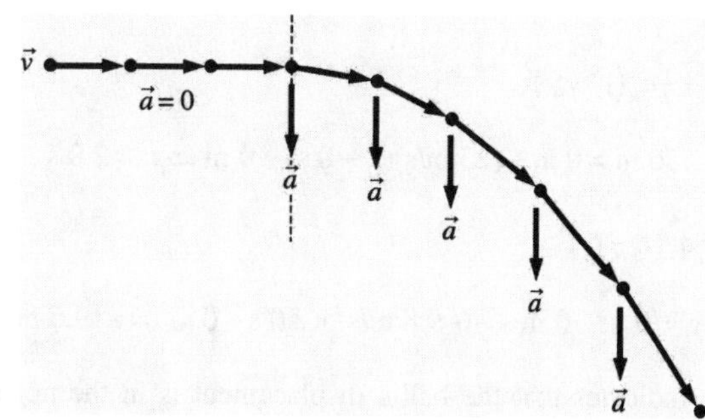


**(b)** A ball rolls along a level table at 3 m/s. It rolls over the edge and falls 1 m to the floor. How far away from the edge of the table does it land?

**4.7. Visualize:** Refer to Figure EX4.7.
**Solve:** From the figure, identify the following:

$$x_1 = 0 \text{ m} \qquad y_1 = 0 \text{ m}$$
$$x_2 = 2000 \text{ m} \qquad y_2 = 1000 \text{ m}$$
$$v_{1x} = 0 \text{ m/s} \qquad v_{1y} = 200 \text{ m/s}$$
$$v_{2x} = 200 \text{ m/s} \qquad v_{2y} = -100 \text{ m/s}$$

The components of the acceleration can be found by applying $v_2^2 = v_1^2 + 2a\Delta s$ for the $x$ and $y$ directions. Thus

$$a_x = \frac{v_{2x}^2 - v_{1x}^2}{2\Delta x} = \frac{(200 \text{ m/s})^2 - (0 \text{ m/s})^2}{2(2000 \text{ m} - 0 \text{ m})} = 10.00 \text{ m/s}^2$$

$$a_y = \frac{(-100 \text{ m/s})^2 - (200 \text{ m/s})^2}{2(1000 \text{ m} - 0 \text{ m})} = -15.00 \text{ m/s}^2$$

So $\vec{a} = (10.00\hat{i} - 15.00\hat{j}) \text{ m/s}^2$.

**Assess:** A time of 20 s is needed to change $v_{1x} = 0$ m/s to $v_{2x} = 200$ m/s at $a_x = 10$ m/s$^2$. This is the same time needed to change $v_{1y}$ to $v_{2y}$ at $a_y = -15$ m/s$^2$.

**4.11. Model:** The ball is treated as a particle and the effect of air resistance is ignored.
**Visualize:**

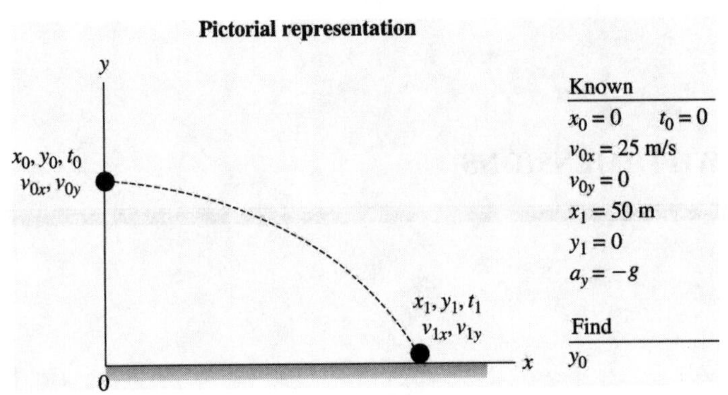


**Solve:** Using $x_1 = x_0 + v_{0x}(t_1 - t_0) + \frac{1}{2}a_x(t_1 - t_0)^2$,

$$50 \text{ m} = 0 \text{ m} + (25 \text{ m/s})(t_1 - 0 \text{ s}) + 0 \text{ m} \Rightarrow t_1 = 2.0 \text{ s}$$

Now, using $y_1 = y_0 + v_{0y}(t_1 - t_0) + \frac{1}{2}a_y(t_1 - t_0)^2$,

$$y_1 = 0 \text{ m} + 0 \text{ m} + \tfrac{1}{2}(-9.8 \text{ m/s}^2)(2.0 \text{ s} - 0 \text{ s})^2 = -19.6 \text{ m}$$

**Assess:** The minus sign with $y_1$ indicates that the ball's displacement is in the negative $y$ direction or downward. A magnitude of 19.6 m for the height is reasonable.

**4.17. Motion:** Assume motion along the $x$-direction. Let the earth frame be S and a frame attached to the moving sidewalk be S′. Frame S′ moves relative to S with velocity $V_x$.
**Solve:** Let $v_x$ be your velocity in frame S and $v_x'$ be your velocity in S′. In the first case, when the moving sidewalk is broken, $V_x = 0$ m/s and

$$v_{x\,\text{W}} = \frac{(x_1 - x_0)}{50 \text{ s}}$$

In the second case, when you stand on the moving sidewalk, $v_x' = 0$ m/s. Therefore, using $v_x = v_x' + V_x$, we get

$$v_{x\,\text{S}} = V_x = \frac{x_1 - x_0}{75 \text{ s}}$$

In the third case, when you walk while riding, $v_x' = v_{x\text{W}}$. Using $v_x = v_x' + V_x$, we get

$$\frac{x_1 - x_0}{t} = \frac{x_1 - x_0}{50 \text{ s}} + \frac{x_1 - x_0}{75 \text{ s}} \Rightarrow t = 30 \text{ s}$$

**Assess:** A time smaller than 50 s was expected.

**4.19. Model:** Let Susan's frame be S and Shawn's frame be S′. S′ moves relative to S with velocity $V$. Both Susan and Shawn are observing the intersection point from their frames.
**Solve:** The Galilean transformation of velocity is $\vec{v} = \vec{v}' + \vec{V}$, where $\vec{v}$ is the velocity of the intersection point from Susan's reference frame, $\vec{v}'$ is the velocity of the intersection point from Shawn's frame S′, and $\vec{V}$ is the velocity of S′ relative to S or Shawn's velocity relative to Susan. Because $\vec{v} = -(60 \text{ mph})\hat{j}$ and $\vec{v}' = -(45 \text{ mph})\hat{i}$, we have $\vec{V} = \vec{v} - \vec{v}' = (45 \text{ mph})\hat{i} - (60 \text{ mph})\hat{j}$. This means that Shawn's speed relative to Susan is

$$V = \sqrt{(45 \text{ mph})^2 + (-60 \text{ mph})^2} = 75 \text{ mph}$$

**4.23. Model:** Treat the record on a turntable as a particle rotating at 45 rpm.
**Solve: (a)** The angular velocity is

$$\omega = 45 \text{ rpm} \times \frac{1 \text{min}}{60 \text{ s}} \times \frac{2\pi \text{ rad}}{1 \text{ rev}} = 1.5\pi \text{ rad/s}$$

**(b)** The period is

$$T = \frac{2\pi \text{ rad}}{|\omega|} = \frac{2\pi \text{ rad}}{1.5\pi \text{ rad/s}} = 1.33 \text{ s}$$

**4.27. Model:** The rider is assumed to be a particle.
**Solve:** Since $a_r = v^2/r$, we have

$$v^2 = a_r r = (98 \text{ m/s}^2)(12 \text{ m}) \Rightarrow v = 34 \text{ m/s}$$

**Assess:** $34 \text{ m/s} \approx 70 \text{ mph}$ is a large yet understandable speed.

**4.35. Model:** Model the child on the merry-go-round as a particle in nonuniform circular motion.
**Visualize:**

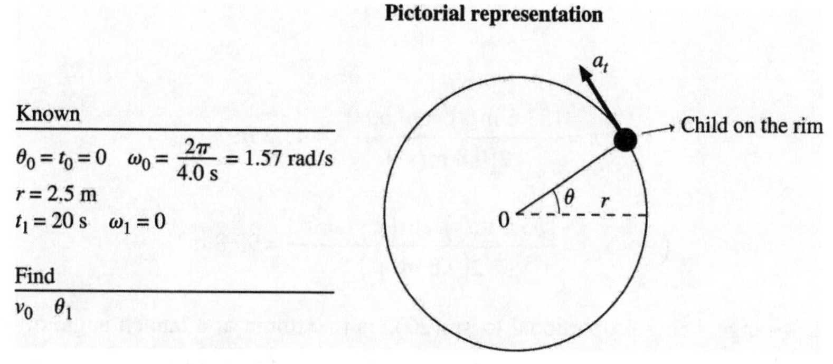


**Solve:** **(a)** The speed of the child is $v_0 = r\omega = (2.5 \text{ m})(1.57 \text{ rad/s}) = 3.9 \text{ m/s}$.
**(b)** The merry-go-round slows from 1.57 rad/s to 0 in 20 s. Thus

$$\omega_1 = 0 = \omega_0 + \frac{a_t}{r} t_1 \Rightarrow a_t = -\frac{r\omega_0}{t_1} = -\frac{(2.5 \text{ m})(1.57 \text{ rad/s})}{20 \text{ s}} = -0.197 \text{ m/s}^2$$

During these 20 s, the wheel turns through angle

$$\theta_1 = \theta_0 + \omega_0 t_1 + \frac{a_t}{2r} t_1^2 = 0 + (1.57 \text{ rad/s})(20 \text{ s}) - \frac{0.197 \text{ m/s}^2}{2(2.5 \text{ m})}(20 \text{ s})^2 = 15.6 \text{ rad}$$

In terms of revolutions, $\theta_1 = (15.6 \text{ rad})(1 \text{ rev}/2\pi\text{rad}) = 2.49 \text{ rev}$.

**4.41. Model:** Assume particle motion in a plane and constant-acceleration kinematics for the projectile.
**Visualize:**

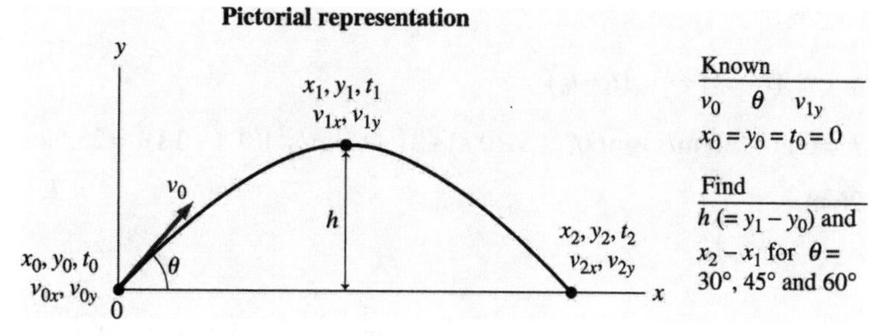

**Solve:** **(a)** We know that $v_{0y} = v_0 \sin\theta$, $a_y = -g$, and $v_{1y} = 0$ m/s. Using $v_{1y}^2 = v_{0y}^2 + 2a_y(y_1 - y_0)$,

$$0 \text{ m}^2/\text{s}^2 = v_0^2 \sin^2\theta + 2(-g)h \Rightarrow h = \frac{v_0^2 \sin^2\theta}{2g}$$

**(b)** Using Equation 4.19 and the above expression for $\theta = 30.0°$:

$$h = \frac{(33.6 \text{ m/s})^2 \sin^2 30.0°}{2(9.8 \text{ m/s}^2)} = 14.4 \text{ m}$$

$$(x_2 - x_0) = \frac{v_0^2 \sin 2\theta}{g} = \frac{(33.6 \text{ m/s})^2 \sin(2\times 30.0°)}{(9.8 \text{ m/s}^2)} = 99.8 \text{ m}$$

For $\theta = 45.0°$:

$$h = \frac{(33.6 \text{ m/s})^2 \sin^2 45.0°}{2(9.8 \text{ m/s}^2)} = 28.8 \text{ m}$$

$$(x_2 - x_0) = \frac{(33.6 \text{ m/s})^2 \sin(2\times 45.0°)}{(9.8 \text{ m/s}^2)} = 115.2 \text{ m}$$

For $\theta = 60.0°$:

$$h = \frac{(33.6 \text{ m/s})^2 \sin^2 60.0°}{2(9.8 \text{ m/s}^2)} = 43.2 \text{ m}$$

$$(x_2 - x_0) = \frac{(33.6 \text{ m/s})^2 \sin(2\times 60.0°)}{2(9.8 \text{ m/s}^2)} = 99.8 \text{ m}$$

**Assess:** The projectile's range, being proportional to $\sin(2\theta)$, is maximum at a launch angle of 45°, but the maximum height reached is proportional to $\sin^2(\theta)$. These dependencies are seen in this problem.

**4.45. Model:** The particle model for the ball and the constant-acceleration equations of motion are assumed.
**Visualize:**

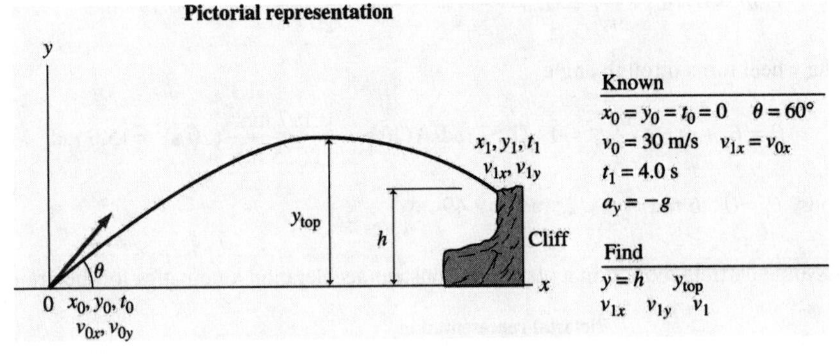


**Solve:** **(a)** Using $y_1 = y_0 + v_{0y}(t_1 - t_0) + \frac{1}{2}a_y(t_1 - t_0)^2$,

$$h = 0 \text{ m} + (30 \text{ m/s})\sin 60°(4 \text{ s} - 0 \text{ s}) + \tfrac{1}{2}(-9.8 \text{ m/s}^2)(4 \text{ s} - 0 \text{ s})^2 = 25.5 \text{ m}$$

The height of the cliff is 26 m.
**(b)** Using $(v_y^2)_{top} = v_y^2 + 2a_y(y_{top} - y_0)$,

$$0 \text{ m}^2/\text{s}^2 = (v_0 \sin\theta)^2 + 2(-g)(y_{top}) \Rightarrow y_{top} = \frac{(v_0 \sin\theta)^2}{2\,g} = \frac{[(30 \text{ m/s})\sin 60°]^2}{2(9.8 \text{ m/s}^2)} = 34.4 \text{ m}$$

The maximum height of the ball is 34 m.
**(c)** The $x$ and $y$ components are

$$v_{1y} = v_{0y} + a_y(t_1 - t_0) = v_0 \sin\theta - gt_1 = (30 \text{ m/s})\sin 60° - (9.8 \text{ m/s}^2)\times(4.0 \text{ s}) = -13.22 \text{ m/s}$$

$$v_{1x} = v_{0y} = v_0 \cos 60° = (30 \text{ m/s})\cos 60° = 15.0 \text{ m/s}$$

$$\Rightarrow v_1 = \sqrt{v_{1x}^2 + v_{1y}^2} = 20.0 \text{ m/s}$$

The impact speed is 20 m/s.
**Assess:** Compared to a maximum height of 34.4 m, a height of 25.5 for the cliff is reasonable.

**4.47. Model:** The particle model for the ball and the constant-acceleration equations of motion in a plane are assumed.
**Visualize:**

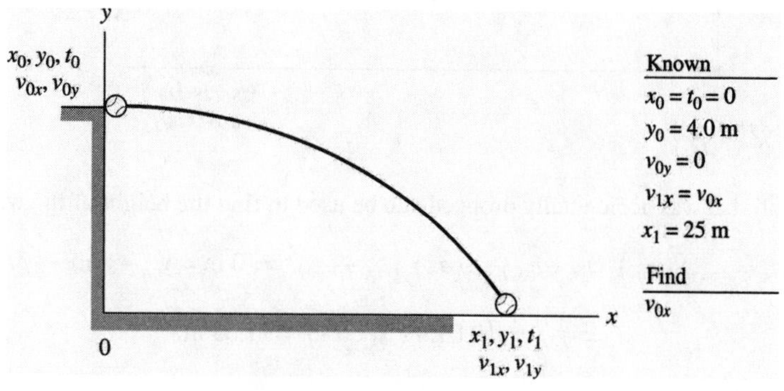


**Solve:** **(a)** The time for the ball to fall is calculated as follows:

$$y_1 = y_0 + v_{0y}(t_1 - t_0) + \tfrac{1}{2}a_y(t_1 - t_0)^2$$

$$\Rightarrow 0 \text{ m} = 4 \text{ m} + 0 \text{ m} + \tfrac{1}{2}(-9.8 \text{ m/s}^2)(t_1 - 0 \text{ s})^2 \Rightarrow t_1 = 0.9035 \text{ s}$$

Using this result for the horizontal velocity:

$$x_1 = x_0 + v_{0x}(t_1 - t_0) \Rightarrow 25 \text{ m} = 0 \text{ m} + v_{0x}(0.9035 \text{ s} - 0 \text{ s}) \Rightarrow v_{0x} = 27.7 \text{ m/s}$$

The friend's pitching speed is 28 m/s.
**(b)** We have $v_{0y} = \pm v_0 \sin\theta$, where we will use the plus sign for up 5° and the minus sign for down 5°. We can write

$$y_1 = y_0 \pm v_0 \sin\theta(t_1 - t_0) - \frac{g}{2}(t_1 - t_0)^2 \Rightarrow 0 \text{ m} = 4 \text{ m} \pm v_0 \sin\theta\, t_1 - \frac{g}{2}t_1^2$$

Let us first find $t_1$ from $x_1 = x_0 + v_{0x}(t_1 - t_0)$:

$$25 \text{ m} = 0 \text{ m} + v_0 \cos\theta\, t_1 \Rightarrow t_1 = \frac{25 \text{ m}}{v_0 \cos\theta}$$

Now substituting $t_1$ into the $y$-equation above yields

$$0 \text{ m} = 4 \text{ m} \pm v_0 \sin\theta\left(\frac{25 \text{ m}}{v_0 \cos\theta}\right) - \frac{g}{2}\left(\frac{25 \text{ m}}{v_0 \cos\theta}\right)^2$$

$$\Rightarrow v_0^2 = \frac{g(25 \text{ m})^2}{2\cos^2\theta}\left\{\frac{1}{4 \text{ m} \pm (25 \text{ m})\tan\theta}\right\} = 22.3 \text{ m/s and } 44.2 \text{ m/s}$$

The range of speeds is 22 m/s to 44 m/s, which is the same as 50 mph to 92 mph.
**Assess:** These are reasonable speeds for baseball pitchers.

**4.51. Model:** We will assume a particle model for the cannonball, and apply the constant-acceleration kinematic equations.
**Visualize:**

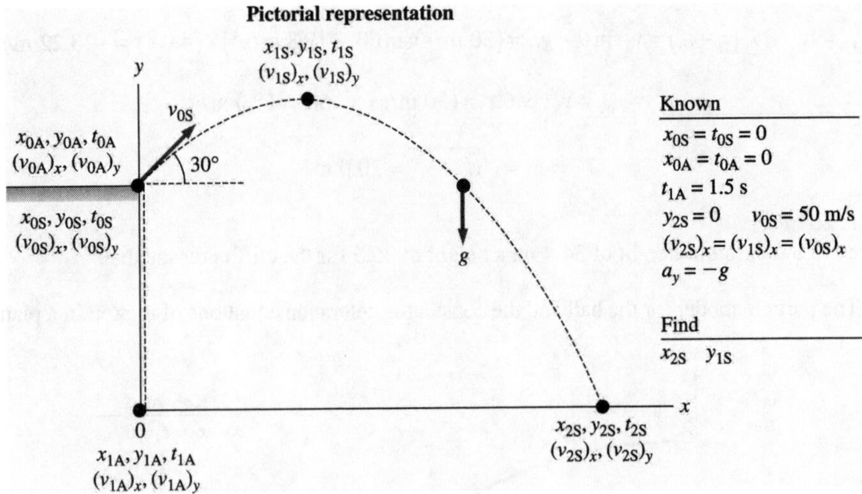


**Solve:** **(a)** The cannonball that was accidentally dropped can be used to find the height of the wall:

$$y_{1A} = y_{0A} + (v_{0A})_y (t_{1A} - t_{0A}) + \tfrac{1}{2}(a_A)_y (t_{1A} - t_{0A})^2 \Rightarrow 0 \text{ m} = y_{0A} + 0 \text{ m} - \tfrac{1}{2} g t_{1A}^2$$

$$\Rightarrow y_{0A} = \tfrac{1}{2}(9.8 \text{ m/s}^2)(1.5 \text{ s})^2 = 11.03 \text{ m}$$

For the cannonball that was shot:

$$(v_{0S})_x = v_{0S} \cos 30° = (50 \text{ m/s}) \cos 30° = 43.30 \text{ m/s}$$

$$(v_{0S})_y = v_{0s} \sin 30° = (50 \text{ m/s}) \sin 30° = 25.0 \text{ m/s}$$

We can now find the time it takes the cannonball to hit the ground:

$$y_{2S} = y_{0S} + (v_{0S})_y (t_{2S} - t_{0S}) + \tfrac{1}{2}(a_S)_y (t_{2S} - t_{0S})^2$$

$$\Rightarrow 0 \text{ m} = (11.03 \text{ m}) + (25.0 \text{ m/s}) t_{2S} - \frac{(9.8 \text{ m/s}^2)}{2} t_{2S}^2$$

$$\Rightarrow (4.9 \text{ m/s}^2) t_{2S}^2 - (25.0 \text{ m/s}) t_{2S} - (11.03 \text{ m}) = 0 \Rightarrow t_{2S} = 5.51 \text{ s}$$

There is also an unphysical root $t_{2S} = -0.41$ s. Using this time $t_{2S}$, we can now find the horizontal distance from the wall as follows:

$$x_{2s} = x_{0s} + (v_{0s})_x (t_{2s} - t_{0s}) = 0 \text{ m} + (43.30 \text{ m/s})(5.51 \text{ s}) = 239 \text{ m}$$

The cannonball hits the ground $2.4 \times 10^2$ m from the castle wall.
**(b)** At the top of the trajectory $(v_{1S})_y = 0$ m/s. Using $(v_{1S})_y^2 = (v_{0S})_y^2 - 2g(y_{1S} - y_{0S})$,

$$0 \text{ m}^2/\text{s}^2 = (25.0 \text{ m/s})^2 - 2(9.8 \text{ m/s}^2)(y_{1s} - 11.03 \text{ m}) \Rightarrow y_{1s} = 42.9 \text{ m}$$

The maximum height above the ground is 43 m.
**Assess:** In view of the fact that the cannonball has a speed of approximately 110 mph, a distance of 239 m for the cannonball to hit the ground is reasonable.

**4.55. Model:** If a frame S′ is in motion with velocity $\vec{V}$ relative to another frame S and has a displacement $\vec{R}$ relative to S, the positions and velocities ($\vec{r}$ and $\vec{v}$) in S are related to the positions and velocities ($\vec{r}'$ and $\vec{v}'$) in S′ as $\vec{r} = \vec{r}' + \vec{R}$ and $\vec{v} = \vec{v}' + \vec{V}$. In the present case, ship A is frame S and ship B is frame S′. Both ships have a common origin at $t = 0$ s. The position and velocity measurements are made in S and S′ relative to their origins.
**Solve:** **(a)** The velocity vectors of the two ships are:

$$\vec{v}_A = (20 \text{ mph})[\cos 30° \hat{i} - \sin 30° \hat{j}] = (17.32 \text{ mph})\hat{i} - (10.0 \text{ mph})\hat{j}$$

$$\vec{v}_B = (25 \text{ mph})[\cos 20° \hat{i} + \sin 20° \hat{j}] = (23.49 \text{ mph})\hat{i} + (8.55 \text{ mph})\hat{j}$$

Since $\vec{r} = \vec{v}\Delta t$,

$$\vec{r}_A = \vec{v}_A(2\text{ h}) = (34.64\text{ miles})\hat{i} - (20.0\text{ miles})\hat{j}$$

$$\vec{r}_B = \vec{v}_B(2\text{ h}) = (46.98\text{ miles})\hat{i} + (17.10\text{ miles})\hat{j}$$

As $\vec{r}_A = \vec{r}_B + \vec{R}$,

$$\vec{R} = \vec{r}_A - \vec{r}_B = (-12.34\text{ miles})\hat{i} - (37.10\text{ miles})\hat{j} \Rightarrow R = 39.1\text{ miles}$$

The distance between the ships two hours after they depart is 39 miles.
**(b)** Because $\vec{v}_A = \vec{v}_B + \vec{V}$,

$$\vec{V} = \vec{v}_A - \vec{v}_B = -(6.17\text{ mph})\hat{i} - (18.55\text{ mph})\hat{j} \Rightarrow V = 19.5\text{ mph}$$

The speed of ship A as seen by ship B is 19.5 mph.
**Assess:** The value of the speed is reasonable.

**4.59. Model:** Let the ground frame be S and the car frame be S′. S′ moves relative to S with a velocity $V$ along the $x$-direction.
**Solve:** The Galilean transformation of velocity is $\vec{v} = \vec{v}' + \vec{V}$ where $\vec{v}$ and $\vec{v}'$ are the velocities of the raindrops in frames S and S′. While driving north, $\vec{V} = (25\text{ m/s})\hat{i}$ and $v = -v_R\cos\theta\,\hat{j} - v_R\sin\theta\,\hat{i}$. Thus,

$$\vec{v}' = \vec{v} - \vec{V} = (-v_R\sin\theta - 25\text{ m/s})\hat{i} - v_R\cos\theta\hat{j}$$

Since the observer in the car finds the raindrops making an angle of 38° with the vertical, we have

$$\frac{v_R\sin\theta + 25\text{ m/s}}{v_R\cos\theta} = \tan 38°$$

While driving south, $\vec{V} = -(25\text{ m/s})\hat{i}$, and $\vec{v} = -v_R\cos\theta\,\hat{j} - v_R\sin\theta\,\hat{i}$. Thus,

$$\vec{v}' = (-v_R\sin\theta + 25\text{ m/s})\hat{i} - v_R\cos\theta\,\hat{j}$$

Since the observer in the car finds the raindrops falling vertically straight, we have

$$\frac{-v_R\sin\theta + 25\text{ m/s}}{v_R\cos\theta} = \tan 0° = 0 \Rightarrow v_R\sin\theta = 25\text{ m/s}$$

Substituting this value of $v_R\sin\theta$ into the expression obtained for driving north yields:

$$\frac{25\text{ m/s} + 25\text{ m/s}}{v_R\cos\theta} = \tan 38° \Rightarrow v_R\cos\theta = \frac{50\text{ m/s}}{\tan 38°} = 64.0\text{ m/s}$$

Therefore, we have for the velocity of the raindrops:

$$(v_R\sin\theta)^2 + (v_R\cos\theta)^2 = (25\text{ m/s})^2 + (64.0\text{ m/s})^2 \Rightarrow v_R^2 = 4721(\text{m/s})^2 \Rightarrow v_R = 68.7\text{ m/s}$$

$$\tan\theta = \frac{v_R\sin\theta}{v_R\cos\theta} = \frac{25\text{ m/s}}{64\text{ m/s}} \Rightarrow \theta = 21.3°$$

The raindrops fall at 689 m/s while making an angle of 21° with the vertical.

**4.63. Model:** Model the car as a particle in nonuniform circular motion.
**Visualize:**

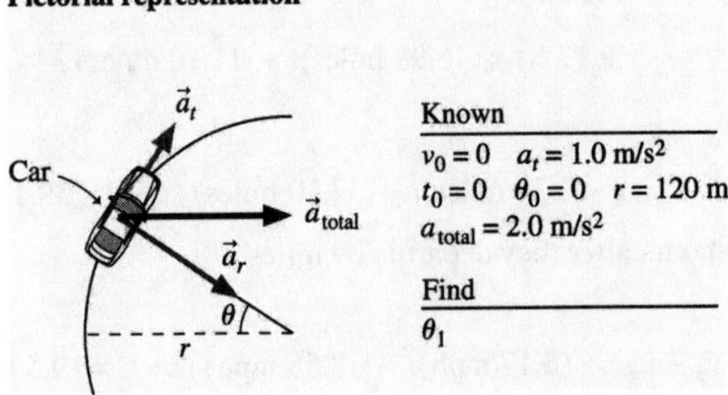


Note that the tangential acceleration stays the same at 1.0 m/s$^2$. As the tangential velocity increases, the radial acceleration increases as well. After a time $t_1$, as the car goes through an angle $\theta_1 - \theta_0$, the total acceleration will increase to 2.0 m/s$^2$. Our objective is to find this angle.

**Solve:** Using $v_1 = v_0 + a_t(t_1 - t_0)$, we get

$$v_1 = 0 \text{ m/s} + (1.0 \text{ m/s}^2)(t_1 - 0 \text{ s}) = (1.0 \text{ m/s}^2)\, t_1$$

$$\Rightarrow a_r = \frac{rv_1^2}{r^2} = \frac{(1.0 \text{ m/s}^2)^2 t_1^2}{120 \text{ m}} = \frac{t_1^2}{120}(\text{m/s}^4)$$

$$\Rightarrow a_{\text{total}} = 2.0 \text{ m/s}^2 = \sqrt{a_t^2 + a_r^2} = \sqrt{(1.0 \text{ m/s}^2)^2 + \left[\frac{t_1^2}{120}(\text{m/s}^4)\right]^2} \Rightarrow t_1 = 14.4 \text{ s}$$

We can now determine the angle $\theta_1$ using

$$\theta_1 = \theta_0 + \omega_0(t_1 - t_0) + \tfrac{1}{2}\left(\frac{a_t}{r}\right)(t_1 - t_0)^2$$

$$= 0 \text{ rad} + 0 \text{ rad} + \frac{1}{2}\frac{(1.0 \text{ m/s}^2)}{(120 \text{ m})}(14.4 \text{ s})^2 = 0.864 \text{ rad} = 49.5°$$

The car will have traveled through an angle of 50°.

**4.67. Model:** The drill is a rigid rotating body.
**Visualize:**

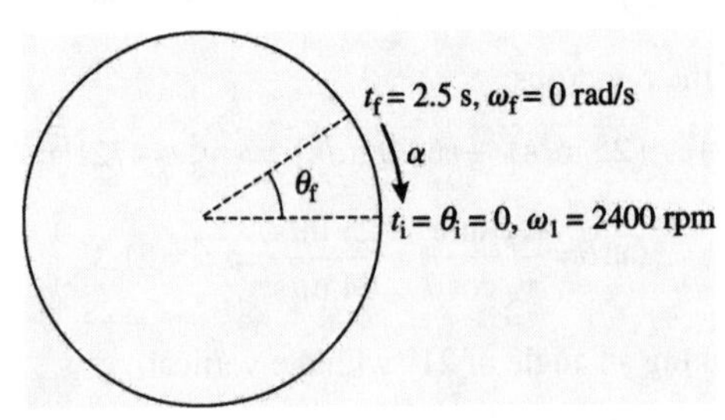


The figure shows the drill's motion from the top.

**Solve:** (a) The kinematic equation $\omega_f = \omega_i + \alpha(t_f - t_i)$ becomes, after using $\omega_i = 2400 \text{ rpm} = (2400)(2\pi)/60 = 251.3$ rad/s, $t_f - t_i = 2.5 \text{ s} - 0 \text{ s} = 2.5$ s, and $w_f = 0$ rad/s,

$$0 \text{ rad} = 251.3 \text{ rad/s} + \alpha(2.5 \text{ s}) \Rightarrow \alpha = -100 \text{ rad/s}^2$$

**(b)** Applying the kinematic equation for angular position yields:

$$\theta_f = \theta_i + \omega_i(t_f - t_i) + \frac{1}{2}\alpha(t_f - t_i)^2$$

$$= 0 \text{ rad} + (251.3 \text{ rad/s})(2.5 \text{ s} - 0 \text{ s}) + \frac{1}{2}(-100 \text{ rad/s}^2)(2.5 \text{ s} - 0 \text{ s})^2$$

$$= 3.2\times 10^2 \text{ rad} = 50 \text{ rev}$$

**4.71. Model:** The bicycle wheel undergoes nonuniform circular motion with constant angular acceleration.
**Visualize:**

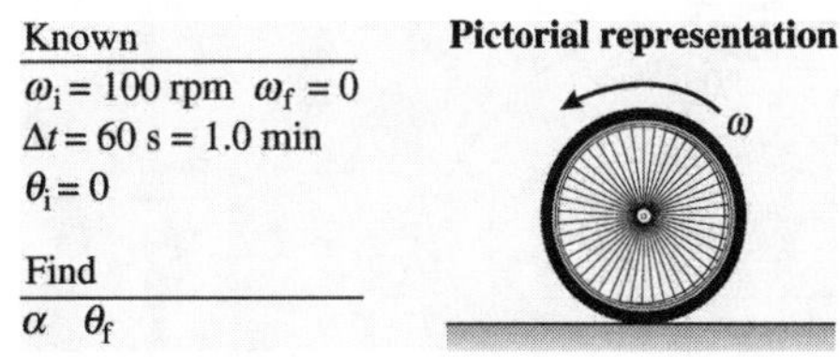


**Solve:** First find the angular acceleration $\alpha$, then use it to find $\theta_f$. Using kinematics,

$$\omega_f = \omega_i + \alpha\Delta t \Rightarrow 0 \text{ rpm} = 100 \text{ rpm} + \alpha(1.0 \text{ min}) \Rightarrow \alpha = -100 \text{ (rpm)/min} = -100 \text{ rev/min}^2$$

The minus sign indicates the wheel is slowing down.
The total number of revolutions the wheel makes while stopping is

$$\theta_f = 0 \text{ rev} + (100 \text{ rpm})(1.0 \text{ min}) + \tfrac{1}{2}(-100 \text{ rev/min}^2)(1.0 \text{ min})^2 = 50 \text{ rev}$$

**Assess:** A total of 50 revolutions in 60 s is on average less than one revolution per second, which is quite reasonable.

**4.75. Solve: (a)** A submarine moving east at 3.0 m/s sees an enemy ship 100 m north of its path. The submarine's torpedo tube happens to be stuck in a position pointing 45° west of north. The tube fires a torpedo with a speed of 6.0 m/s relative to the submarine. How far east or west of the ship should the sub be when it fires?
**(b)** Relative to the water, the torpedo will have velocity components

$$v_x = -6.0\cos 45° \text{ m/s} + 3.0 \text{ m/s} = -4.24 \text{ m/s} + 3 \text{ m/s} = -1.24 \text{ m/s}$$

$$v_y = +6.0\cos 45° \text{ m/s} = +4.2 \text{ m/s}$$

The time to travel north to the ship is

$$100 \text{ m} = (4.2 \text{ m/s})\ t_1 \Rightarrow t_1 = 24 \text{ s}$$

Thus, $x = (1.24 \text{ m/s})(24 \text{ s}) = -30$ m. That is, the ship should be 30 m west of the submarine.

**4.77. Solve:** You decide to test fly your model airplane off of a 125 m tall building. The model's engine starts fine and gets the airplane moving at 4.0 m/s but quits just as it gets to the edge of the building. The model proceeds to fall "like a rock." How far from the edge of the building will it crash into the ground? (Assume $g = 10\ \text{m/s}^2$ for easier calculation.)

**Visualize:**

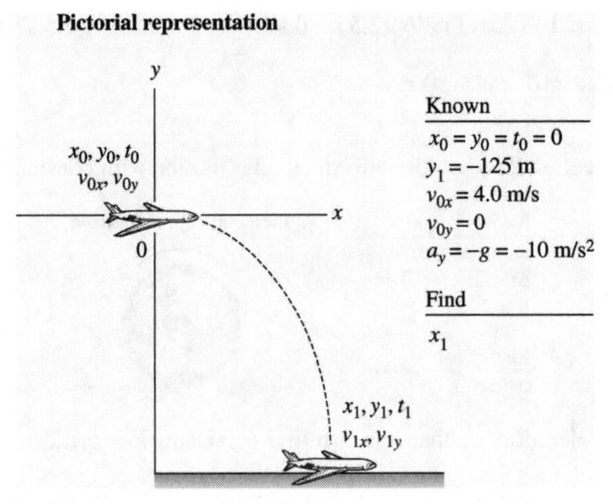


Using the equation $y_1 = y_0 + v_{0y}t + \frac{1}{2}a_y(t_0 - t_1)^2$, we get

$$y_1 = -(5\ \text{m/s}^2)\,t_1^2 = -125\ \text{m} \Rightarrow t_1 = \sqrt{\frac{125\ \text{m}}{5\ \text{m/s}^2}} = \sqrt{25\ \text{s}^2} = 5\ \text{s}$$

The distance $x_1 = (4\ \text{m/s})(5\ \text{s}) = 20$ m.

# 5

## FORCE AND MOTION

**5.9. Visualize:** Please refer to Figure EX5.9.
**Solve:** Mass is defined to be

$$m = \frac{1}{\text{slope of the acceleration-versus-force graph}}$$

Thus

$$\frac{m_1}{m_2} = \frac{(\text{slope of line 1})^{-1}}{(\text{slope of line 2})^{-1}} = \frac{\left(\frac{5a_1}{3N}\right)^{-1}}{\left(\frac{3a_1}{5N}\right)^{-1}} = \frac{\frac{3}{5}}{\frac{5}{3}} = \frac{9}{25}$$

The ratio of masses is

$$\frac{m_1}{m_2} = \frac{9}{25}$$

**Assess:** More rubber bands produce a smaller acceleration on object 2, so it should be more massive.

**5.15. Visualize:** Please refer to Figure EX5.15.
**Solve:** Newton's second law is $F = ma$. We can read a force and an acceleration from the graph, and hence find the mass. Choosing the force $F = 1$ N gives us $a = 4$ m/s$^2$. Newton's second law yields $m = 0.25$ kg.

**5.19. Visualize:**

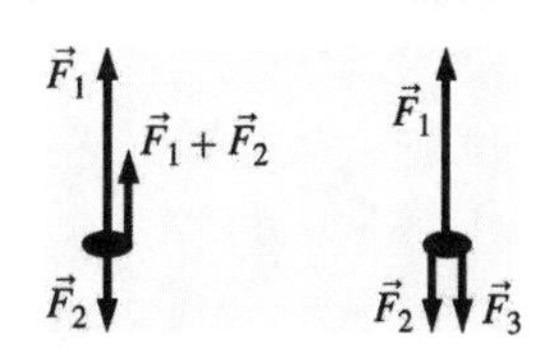


**Solve:** The object will be in equilibrium if $\vec{F}_3$ has the same magnitude as $\vec{F}_1 + \vec{F}_2$ but is in the opposite direction so that the sum of all three forces is zero.

**5.25. Visualize:**

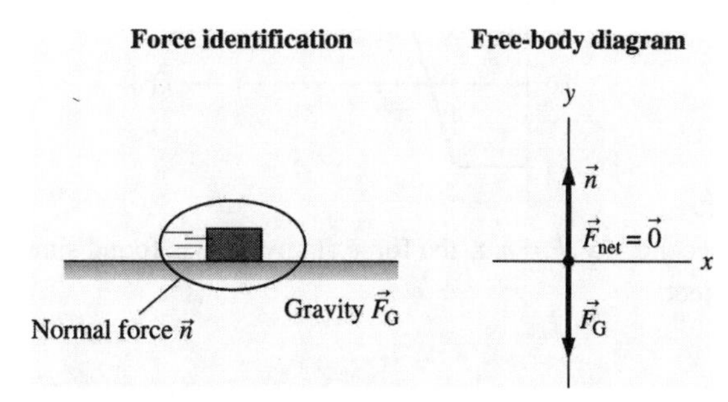


**Assess:** The problem says that there is no friction and it tells you nothing about any drag; so we do not include either of these forces. The only remaining forces are the weight and the normal force.

**5.27. Visualize:**

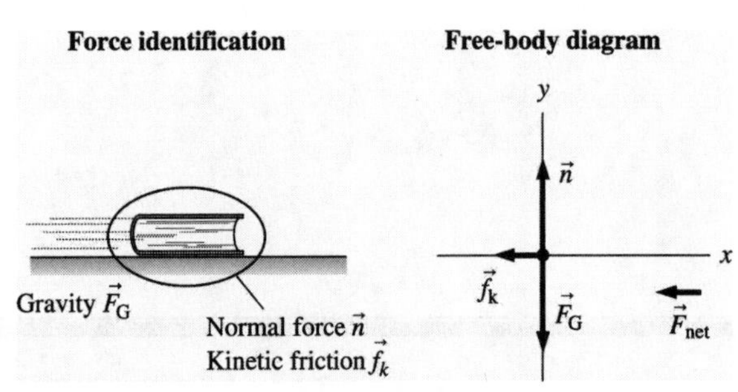


**Assess:** The problem uses the word "sliding." Any real situation involves friction with the surface. Since we are not told to neglect it, we show that force.

**5.29. Visualize:**

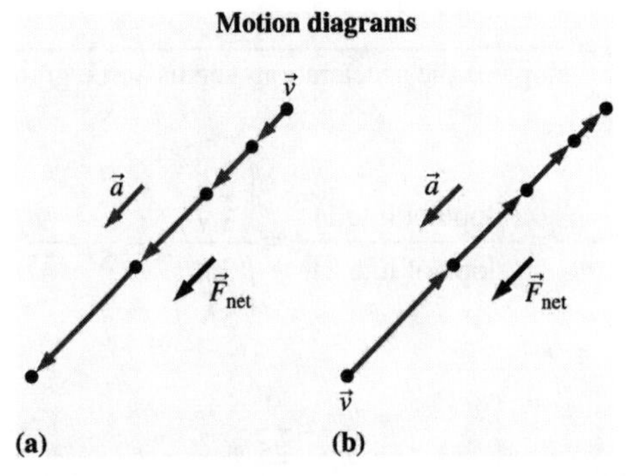


The velocity vector in figure (a) is shown downward and to the left. So movement is downward and to the left. The velocity vectors get successively longer, which means the speed is increasing. Therefore the acceleration is downward and to the left. By Newton's second law $\vec{F} = m\vec{a}$, the net force must be in the same direction as the acceleration. Thus, the net force is downward and to the left.
The velocity vector in (b) is shown to be upward and to the right. So movement is upward and to the right. The velocity vector gets successively shorter, which means the speed is decreasing. Therefore the acceleration is downward and to the left. From Newton's second law, the net force must be in the direction of the acceleration and so it is directed downward and to the left.

**5.31. Visualize:**

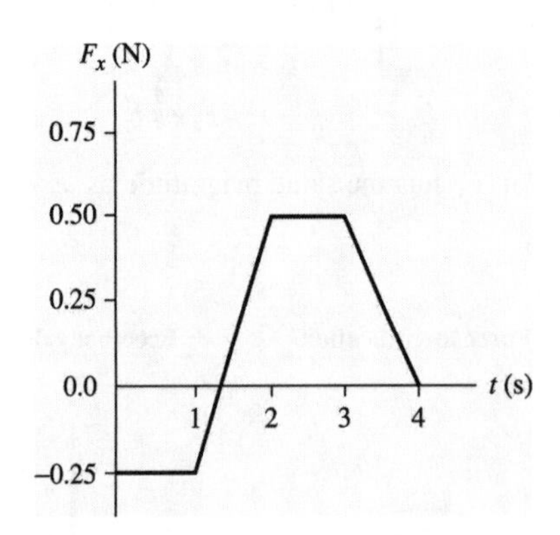


**Solve:** According to Newton's second law $F = ma$, the force at any time is found simply by multiplying the value of the acceleration by the mass of the object.

**5.33. Visualize:**

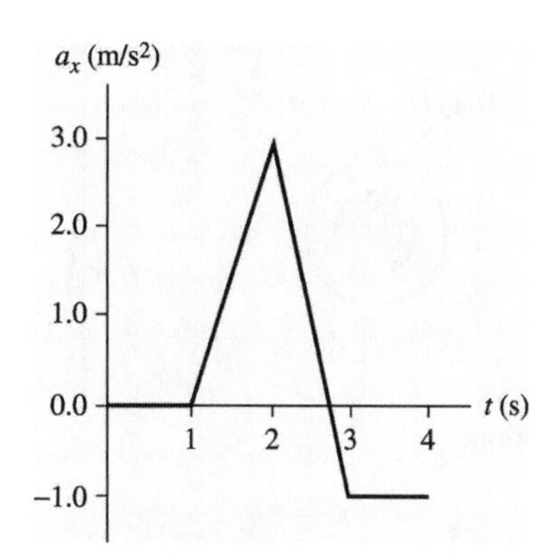


**Solve:** According to Newton's second law $F = ma$, the acceleration at any time is found simply by dividing the value of the force by the mass of the object.

**5.43. Visualize:**

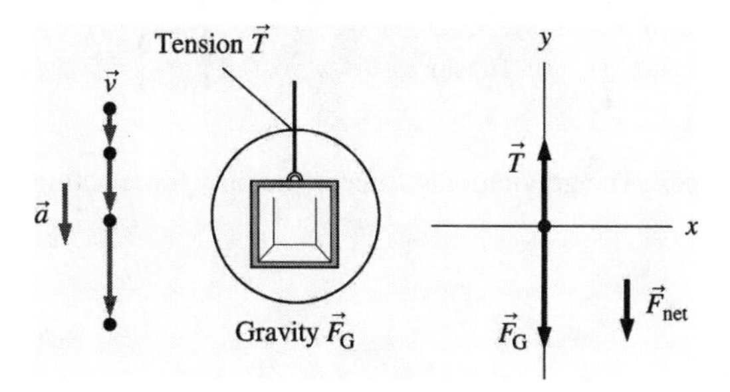


Tension is the only contact force. The downward acceleration implies that $F_G > T$.

**5.45. Visualize:**

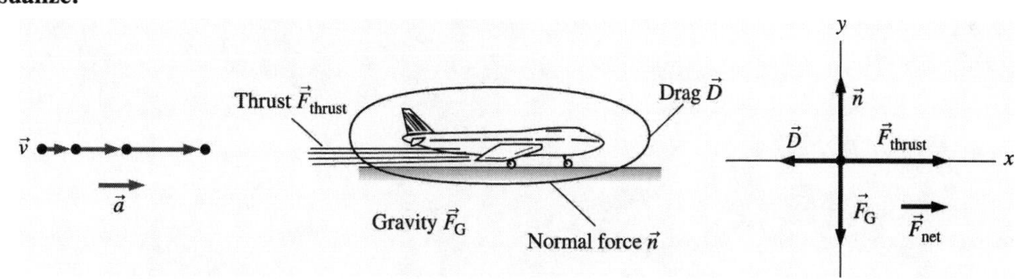


The normal force is perpendicular to the ground. The thrust force is parallel to the ground and in the direction of acceleration. The drag force is opposite to the direction of motion.

**5.47. Visualize:**

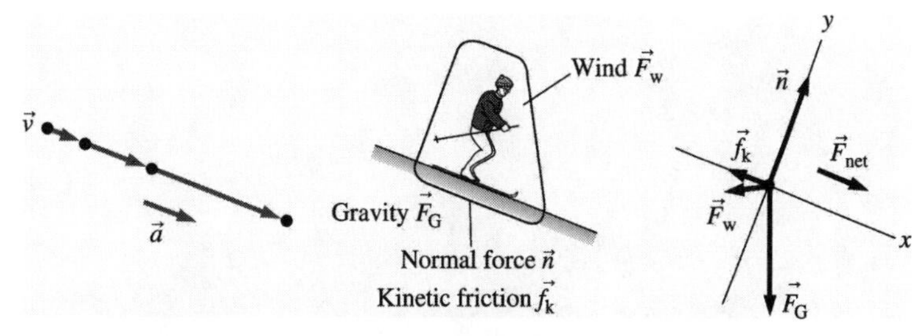


The normal force is perpendicular to the hill. The kinetic frictional force is parallel to the hill and directed upward opposite to the direction of motion. The wind force is given as *horizontal.* Since the skier stays on the slope (that is, there is no acceleration away from the slope) the net force must be parallel to the slope.

**5.49. Visualize:**

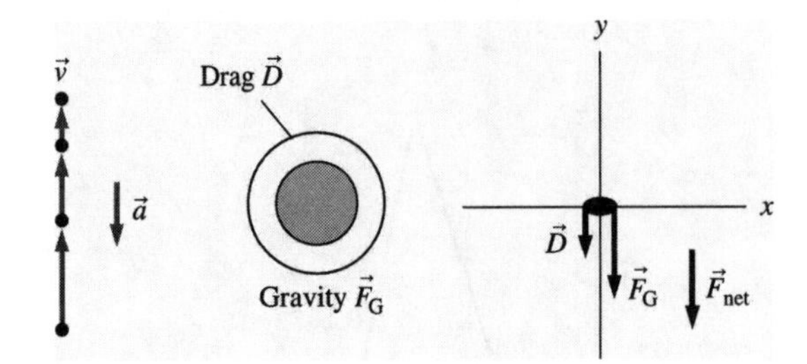


The drag force due to air is opposite the motion.

**5.51. Visualize:**

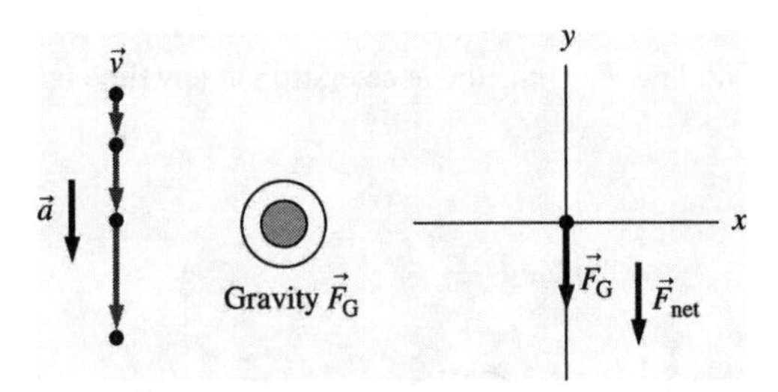


There are no contact forces on the rock. The gravitational force is the only force acting on the rock.

# 6

# DYNAMICS I: MOTION ALONG A LINE

**6.1. Model:** We can assume that the ring is a single massless particle in static equilibrium.
**Visualize:**

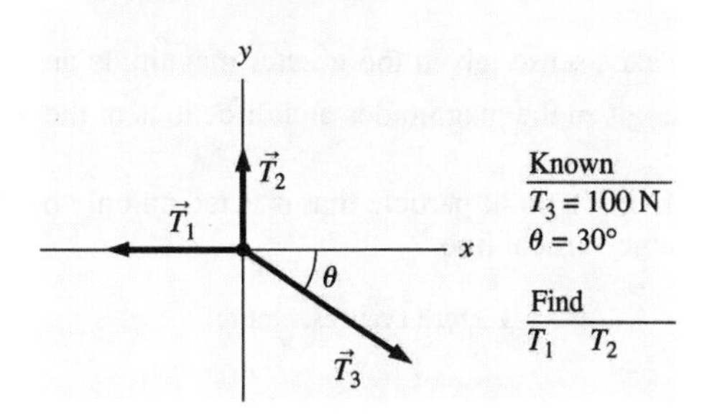


**Solve:** Written in component form, Newton's first law is

$$(F_{net})_x = \Sigma F_x = T_{1x} + T_{2x} + T_{3x} = 0 \text{ N} \quad (F_{net})_y = \Sigma F_y = T_{1y} + T_{2y} + T_{3y} = 0 \text{ N}$$

Evaluating the components of the force vectors from the free-body diagram:

$$T_{1x} = -T_1 \quad T_{2x} = 0 \text{ N} \quad T_{3x} = T_3 \cos 30°$$

$$T_{1y} = 0 \text{ N} \quad T_{2y} = T_2 \quad T_{3y} = -T_3 \sin 30°$$

Using Newton's first law:

$$-T_1 + T_3 \cos 30° = 0 \text{ N} \quad T_2 - T_3 \sin 30° = 0 \text{ N}$$

Rearranging:

$$T_1 = T_3 \cos 30° = (100 \text{ N})(0.8666) = 86.7 \text{ N} \quad T_2 = T_3 \sin 30° = (100 \text{ N})(0.5) = 50.0 \text{ N}$$

**Assess:** Since $\vec{T}_3$ acts closer to the $x$-axis than to the $y$-axis, it makes sense that $T_1 > T_2$.

**6.5. Visualize:** Please refer to the Figure EX6.5.
**Solve:** Applying Newton's second law to the diagram on the left,

$$a_x = \frac{(F_{net})_x}{m} = \frac{4 \text{ N} - 2 \text{ N}}{2 \text{ kg}} = 1.0 \text{ m/s}^2 \qquad a_y = \frac{(F_{net})_y}{m} = \frac{3 \text{ N} - 3 \text{ N}}{2 \text{ kg}} = 0 \text{ m/s}^2$$

For the diagram on the right:

$$a_x = \frac{(F_{net})_x}{m} = \frac{4 \text{ N} - 2 \text{ N}}{2 \text{ kg}} = 1.0 \text{ m/s}^2 \qquad a_y = \frac{(F_{net})_y}{m} = \frac{3 \text{ N} - 1 \text{ N} - 2 \text{ N}}{2 \text{ kg}} = 0 \text{ m/s}^2$$

**6.9. Visualize:** Please refer to Figure EX6.9. Positive forces result in the object gaining speed and negative forces result in the object slowing down. The final segment of zero force is a period of constant speed.

**Solve:** We have the mass and net force for all the three segments. This means we can use Newton's second law to calculate the accelerations. The acceleration from $t = 0$ s to $t = 3$ s is

$$a_x = \frac{F_x}{m} = \frac{4 \text{ N}}{2.0 \text{ kg}} = 2 \text{ m/s}^2$$

The acceleration from $t = 3$ s to $t = 5$ s is

$$a_x = \frac{F_x}{m} = \frac{-2 \text{ N}}{2.0 \text{ kg}} = -1 \text{ m/s}^2$$

The acceleration from $t = 5$ s to 8 s is $a_x = 0 \text{ m/s}^2$. In particular, $a_x(\text{at } t = 6 \text{ s}) = 0 \text{ m/s}^2$.

We can now use one-dimensional kinematics to calculate $v$ at $t = 6$ s as follows:

$$v = v_0 + a_1(t_1 - t_0) + a_2(t_2 - t_0)$$
$$= 0 + (2 \text{ m/s}^2)(3 \text{ s}) + (-1 \text{ m/s}^2)(2 \text{ s}) = 6 \text{ m/s} - 2 \text{ m/s} = 4 \text{ m/s}$$

**Assess:** The positive final velocity makes sense, given the greater magnitude and longer duration of the positive $\vec{F}_1$. A velocity of 4 m/s also seems reasonable, given the magnitudes and directions of the forces and the mass involved.

**6.11. Model:** We assume that the box is a point particle that is acted on only by the tension in the rope and the pull of gravity. Both the forces act along the same vertical line.

**Visualize:**

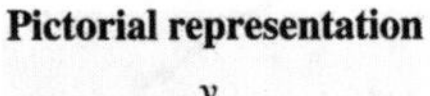


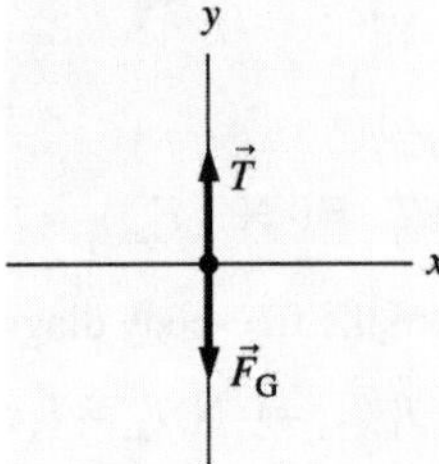


**Solve:** **(a)** Since the box is at rest, $a_y = 0 \text{ m/s}^2$ and the net force on it must be zero:

$$F_{net} = T - F_G = 0 \text{ N} \Rightarrow T = F_G = mg = (50 \text{ kg})(9.8 \text{ m/s}^2) = 490 \text{ N}$$

**(b)** Since the box is rising at a constant speed, again $a_y = 0 \text{ m/s}^2$, $F_{net} = 0$ N, and $T = F_G = 490$ N.

**(c)** The velocity of the box is irrelevant, since only a *change* in velocity requires a nonzero net force. Since $a_y = 5.0 \text{ m/s}^2$,

$$F_{net} = T - F_G = ma_y = (50 \text{ kg})(5.0 \text{ m/s}^2) = 250 \text{ N}$$
$$\Rightarrow T = 250 \text{ N} + w = 250 \text{ N} + 490 \text{ N} = 740 \text{ N}$$

**(d)** The situation is the same as in part (c), except that the rising box is slowing down. Thus $a_y = -5.0 \text{ m/s}^2$ and we have instead

$$F_{net} = T - F_G = ma_y = (50 \text{ kg})(-5.0 \text{ m/s}^2) = -250 \text{ N}$$
$$\Rightarrow T = -250 \text{ N} + F_G = -250 \text{ N} + 490 \text{ N} = 240 \text{ N}$$

**Assess:** For parts (a) and (b) the zero accelerations immediately imply that the gravitational force on the box must be exactly balanced by the upward tension in the rope. For part (c) the tension not only has to support the gravitational force on the box but must also accelerate it upward, hence, $T$ must be greater than $F_G$. When the box accelerates downward, the rope need not support the entire gravitational force, hence, $T$ is less than $F_G$.

**6.15. Model:** We assume that the passenger is a particle subject to two vertical forces: the downward pull of gravity and the upward push of the elevator floor. We can use one-dimensional kinematics and Equation 6.10.

**Visualize:**

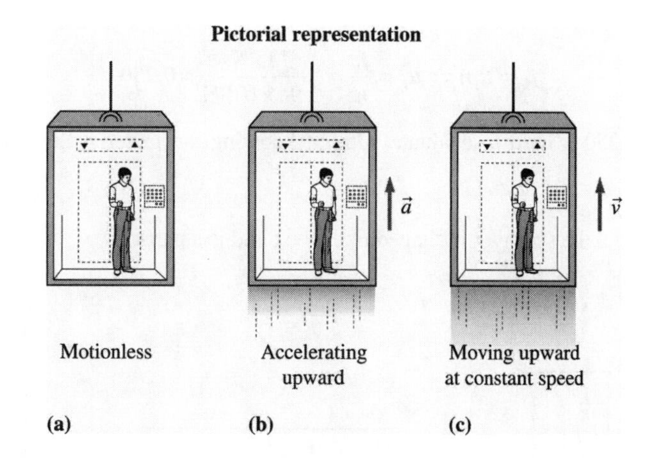


**Solve:** **(a)** The weight is

$$w = mg\left(1+\frac{a_y}{g}\right) = mg\left(1+\frac{0}{g}\right) = mg = (60\text{ kg})(9.80\text{ m/s}^2) = 590\text{ N}$$

**(b)** The elevator speeds up from $v_{0y} = 0$ m/s to its cruising speed at $v_y = 10$ m/s. We need its acceleration before we can find the apparent weight:

$$a_y = \frac{\Delta v}{\Delta t} = \frac{10\text{ m/s} - 0\text{ m/s}}{4.0\text{ s}} = 2.5\text{ m/s}^2$$

The passenger's weight is

$$w = mg\left(1+\frac{a_y}{g}\right) = (590\text{ N})\left(1+\frac{2.5\text{ m/s}^2}{9.80\text{ m/s}^2}\right) = (590\text{ N})(1.26) = 740\text{ N}$$

**(c)** The passenger is no longer accelerating since the elevator has reached its cruising speed. Thus, $w = mg = 590$ N as in part (a).

**Assess:** The passenger's weight is the gravitational force on the passenger in parts (a) and (c), since there is no acceleration. In part (b), the elevator must not only support the gravitational force but must also accelerate him upward, so it's reasonable that the floor will have to push up harder on him, increasing his weight.

**6.17. Model:** We assume that the safe is a particle moving only in the *x*-direction. Since it is sliding during the entire problem, we can use the model of kinetic friction.

**Visualize:**

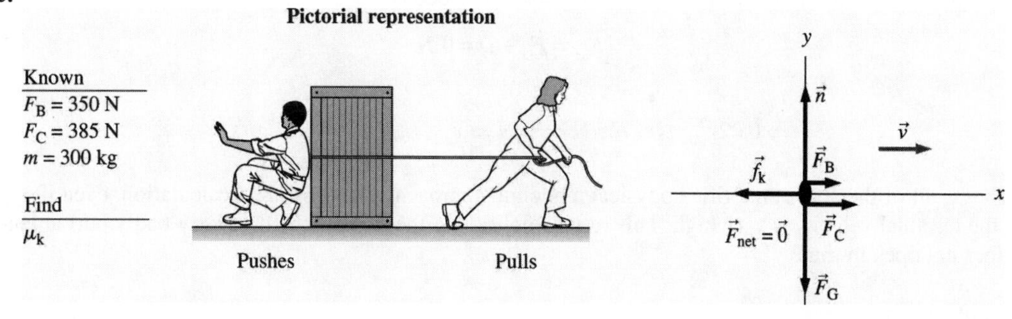


**Solve:** The safe is in equilibrium, since it's not accelerating. Thus we can apply Newton's first law in the vertical and horizontal directions:

$$(F_{net})_x = \Sigma F_x = F_B + F_C - f_k = 0\text{ N} \Rightarrow f_k = F_B + F_C = 350\text{ N} + 385\text{ N} = 735\text{ N}$$

$$(F_{net})_y = \Sigma F_y = n - F_G = 0\text{ N} \Rightarrow n = F_G = mg = (300\text{ kg})(9.80\text{ m/s}^2) = 2.94\times10^3\text{ N}$$

Then, for kinetic friction:

$$f_k = \mu_k n \Rightarrow \mu_k = \frac{f_k}{n} = \frac{735 \text{ N}}{2.94 \times 10^3 \text{ N}} = 0.250$$

**Assess:** The value of $\mu_k = 0.250$ is hard to evaluate without knowing the material the floor is made of, but it seems reasonable.

**6.25. Model:** We assume that the skydiver is shaped like a box and is a particle.
**Visualize:**

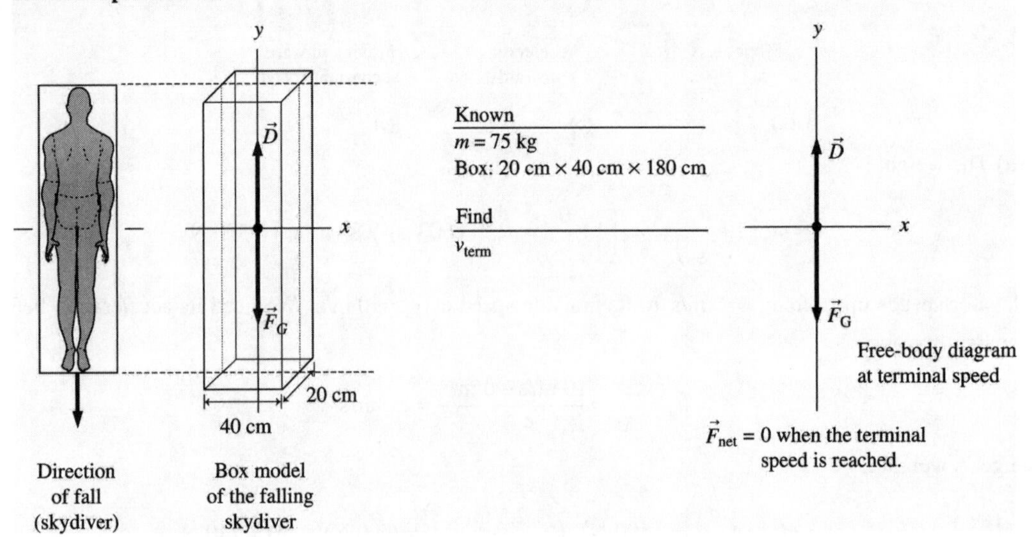


The skydiver falls straight down toward the earth's surface, that is, the direction of fall is vertical. Since the skydiver falls feet first, the surface perpendicular to the drag has the cross-sectional area $A = 20 \text{ cm} \times 40 \text{ cm}$. The physical conditions needed to use Equation 6.16 for the drag force are satisfied. The terminal speed corresponds to the situation when the net force acting on the skydiver becomes zero.
**Solve:** The expression for the magnitude of the drag with $v$ in m/s is

$$D \approx \frac{1}{4}Av^2 = 0.25(0.20 \times 0.40)v^2 \text{ N} = 0.020v^2 \text{ N}$$

The gravitational force on the skydiver is $F_G = mg = (75 \text{ kg})(9.8 \text{ m/s}^2) = 735 \text{ N}$. The mathematical form of the condition defining dynamical equilibrium for the skydiver and the terminal speed is

$$\vec{F}_{net} = \vec{F}_G + \vec{D} = 0 \text{ N}$$

$$\Rightarrow 0.02v_{term}^2 \text{ N} - 735 \text{ N} = 0 \text{ N} \Rightarrow v_{term} = \sqrt{\frac{735}{0.02}} \approx 192 \text{ m/s}$$

**Assess:** The result of the above simplified physical modeling approach and subsequent calculation, even if approximate, shows that the terminal velocity is very high. This result implies that the skydiver will be very badly hurt at landing if the parachute does not open in time.

**6.27. Visualize:**

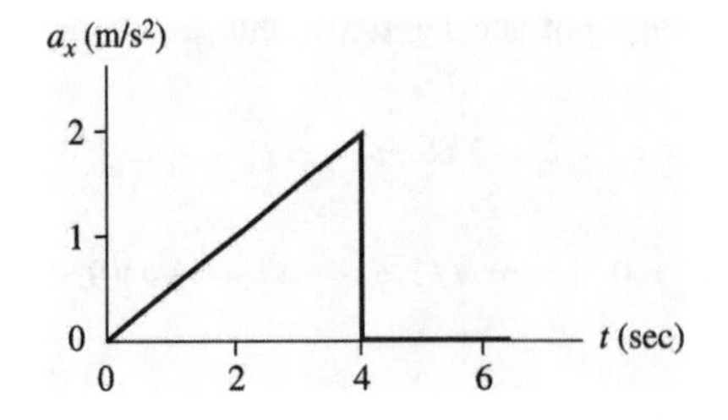

We used the force-versus-time graph to draw the acceleration-versus-time graph. The peak acceleration was calculated as follows:

$$a_{\max} = \frac{F_{\max}}{m} = \frac{10\ \text{N}}{5\ \text{kg}} = 2\ \text{m/s}^2$$

**Solve:** The acceleration is not constant, so we cannot use constant acceleration kinematics. Instead, we use the more general result that

$$v(t) = v_0 + \text{area under the acceleration curve from 0 s to } t$$

The object starts from rest, so $v_0 = 0$ m/s. The area under the acceleration curve between 0 s and 6 s is $\frac{1}{2}(4\ \text{s})(2\ \text{m/s}^2) = 4.0$ m/s. We've used the fact that the area between 4 s and 6 s is zero. Thus, at $t = 6$ s, $v_x = 4.0$ m/s.

**6.33. Model:** We'll assume Zach is a particle moving under the effect of two forces acting in a single vertical line: gravity and the supporting force of the elevator.
**Visualize:**

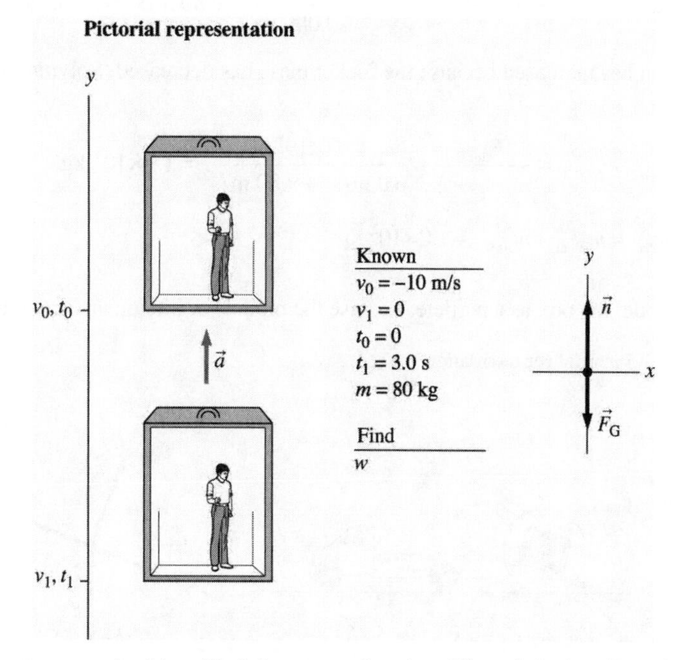


**Solve:** **(a)** Before the elevator starts braking, Zach is not accelerating. His weight (see Equation 6.10) is

$$w = mg\left(1 + \frac{a}{g}\right) = mg\left(1 + \frac{0\ \text{m/s}^2}{g}\right) = mg = (80\ \text{kg})(9.80\ \text{m/s}^2) = 784\ \text{N}$$

Zach's weight is $7.8 \times 10^2$ N.
**(b)** Using the definition of acceleration,

$$a = \frac{\Delta v}{\Delta t} = \frac{v_1 - v_0}{t_1 - t_0} = \frac{0 - (-10)\ \text{m/s}}{3.0\ \text{s}} = 3.33\ \text{m/s}^2$$

$$\Rightarrow w = mg\left(1 + \frac{a}{g}\right) = (80\ \text{kg})(9.80\ \text{m/s}^2)\left(1 + \frac{3.33\ \text{m/s}^2}{9.80\ \text{m/s}^2}\right) = (784\ \text{N})(1 + 0.340) = 1050\ \text{N}$$

Now Zach's weight is $1.05 \times 10^3$ N.
**Assess:** While the elevator is braking, it not only must support the gravitational force on Zach but must also push upward on him to decelerate him, so his weight is greater than the gravitational force.

**6.39. Model:** Represent the rocket as a particle that follows Newton's second law.
**Visualize:**

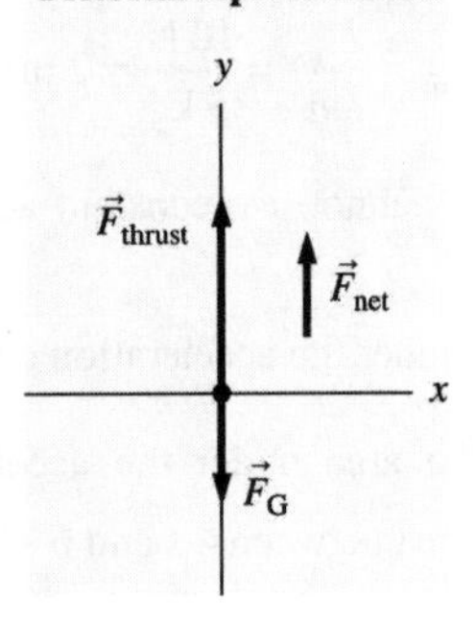


**Solve:** **(a)** The $y$-component of Newton's second law is

$$a_y = a = \frac{(F_{net})_y}{m} = \frac{F_{thrust} - mg}{m} = \frac{3.0\times10^5 \text{ N}}{20{,}000 \text{ kg}} - 9.80 \text{ m/s}^2 = 5.2 \text{ m/s}^2$$

**(b)** At 5000 m the acceleration has increased because the rocket mass has decreased. Solving the equation of part (a) for $m$ gives

$$m_{5000 \text{ m}} = \frac{F_{thrust}}{a_{5000 \text{ m}} + g} = \frac{3.0\times10^5 \text{ N}}{6.0 \text{ m/s}^2 + 9.80 \text{ m/s}^2} = 1.9\times10^4 \text{ kg}$$

The mass of fuel burned is $m_{fuel} = m_{initial} - m_{5000_m} = 1.0\times10^3$ kg.

**6.45. Model:** We will model the box as a particle, and use the models of kinetic and static friction.
**Visualize:**

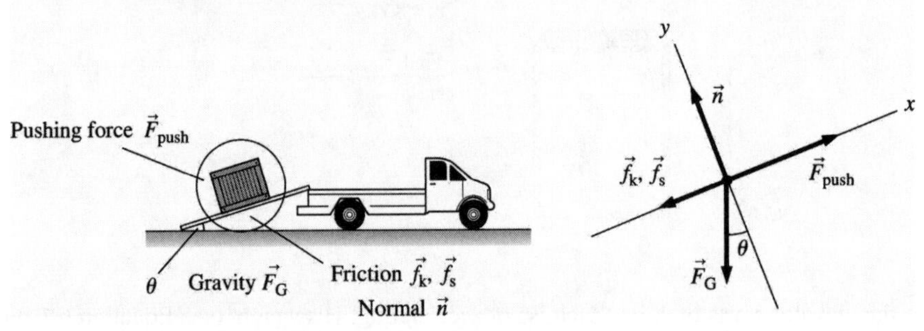


The pushing force is along the +$x$-axis, but the force of friction acts along the –$x$-axis. A component of the gravitational force on the box acts along the –$x$-axis as well. The box will move up if the pushing force is at least equal to the sum of the friction force and the component of the gravitational force in the $x$-direction.
**Solve:** Let's determine how much pushing force you would need to keep the box moving up the ramp at steady speed. Newton's second law for the box in dynamic equilibrium is

$$(F_{net})_x = \Sigma F_x = n_x + (F_G)_x + (f_k)_x + (F_{push})_x = 0 \text{ N} - mg \sin\theta - f_k + F_{push} = 0 \text{ N}$$

$$(F_{net})_y = \Sigma F_y = n_y + (F_G)_y + (f_k)_y + (F_{push})_y = n - mg \cos\theta + 0 \text{ N} + 0 \text{ N} = 0 \text{ N}$$

The $x$-component equation and the model of kinetic friction yield:

$$F_{push} = mg \sin\theta + f_k = mg \sin\theta + \mu_k n$$

Let us obtain $n$ from the $y$-component equation as $n = mg\cos\theta$, and substitute it in the above equation to get

$$F_{push} = mg \sin\theta + \mu_k \, mg \cos\theta = mg(\sin\theta + \mu_k \cos\theta)$$

$$= (100 \text{ kg})(9.80 \text{ m/s}^2)(\sin 20° + 0.60 \cos 20°) = 888 \text{ N}$$

The force is less than your maximum pushing force of 1000 N. That is, once in motion, the box could be kept moving up the ramp. However, if you stop on the ramp and want to start the box from rest, the model of static friction applies. The analysis is the same except that the coefficient of static friction is used and we use the maximum value of the force of static friction. Therefore, we have

$$F_{\text{push}} = mg(\sin\theta + \mu_s \cos\theta) = (100\text{ kg})(9.80\text{ m/s}^2)(\sin 20° + 0.90\cos 20°) = 1160\text{ N}$$

Since you can push with a force of only 1000 N, you can't get the box started. The big static friction force and the weight are too much to overcome.

**6.51. Model:** The box will be treated as a particle. Because the box slides down a vertical wood wall, we will also use the model of kinetic friction.
**Visualize:**

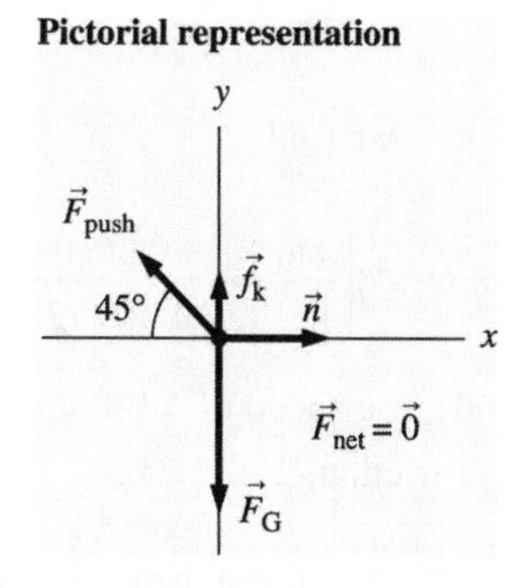


**Solve:** The normal force due to the wall, which is perpendicular to the wall, is here to the right. The box slides down the wall at constant speed, so $\vec{a} = \vec{0}$ and the box is in dynamic equilibrium. Thus, $\vec{F}_{\text{net}} = \vec{0}$. Newton's second law for this equilibrium situation is

$$(F_{\text{net}})_x = 0\text{ N} = n - F_{\text{push}}\cos 45°$$

$$(F_{\text{net}})_y = 0\text{ N} = f_k + F_{\text{push}}\sin 45° - F_G = f_k + F_{\text{push}}\sin 45° - mg$$

The friction force is $f_k = \mu_k n$. Using the $x$-equation to get an expression for $n$, we see that $f_k = \mu_k F_{\text{push}}\cos 45°$. Substituting this into the $y$-equation and using Table 6.1 to find $\mu_k = 0.20$ gives,

$$\mu_k F_{\text{push}}\cos 45° + F_{\text{push}}\sin 45° - mg = 0\text{ N}$$

$$\Rightarrow F_{\text{push}} = \frac{mg}{\mu_k\cos 45° + \sin 45°} = \frac{(2.0\text{ kg})(9.80\text{ m/s}^2)}{0.20\cos 45° + \sin 45°} = 23\text{ N}$$

**6.53. Model:** We will model the skier along with the wooden skis as a particle of mass $m$. The snow exerts a contact force and the wind exerts a drag force on the skier. We will therefore use the models of kinetic friction and drag.
**Visualize:**

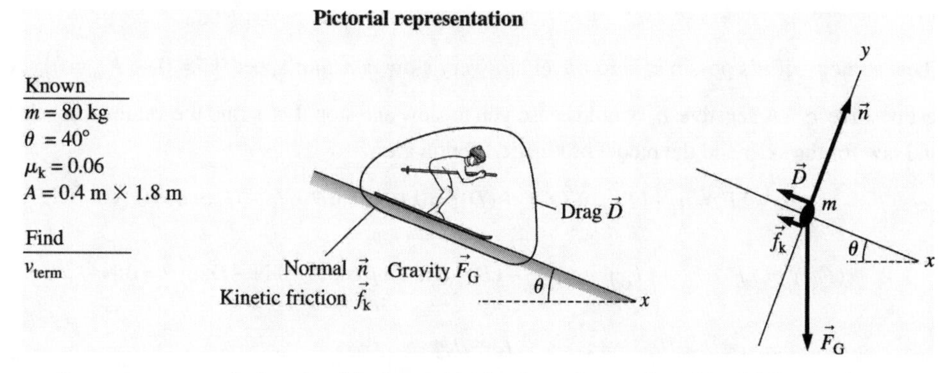


We choose a coordinate system such that the skier's motion is along the +$x$-direction. While the forces of kinetic friction $\vec{f}_k$ and drag $\vec{D}$ act along the –$x$-direction opposing the motion of the skier, the gravitational force on the skier has a component in the +$x$-direction. At the terminal speed, the net force on the skier is zero as the forces along the +$x$-direction cancel out the forces along the –$x$-direction.

**Solve:** Newton's second law and the models of kinetic friction and drag are

$$(F_{\text{net}})_x = \Sigma F_x = n_x + (F_G)_x + (f_k)_x + (D)_x = 0\text{ N} + mg\sin\theta - f_k \frac{1}{4}Av^2 = ma_x = 0\text{ N}$$

$$(F_{\text{net}})_y = \Sigma F_y = n_y + (F_G)_y + (f_k)_y + (D)_y = n - mg\cos\theta + 0\text{ N} + 0\text{ N} = 0\text{ N}$$

$$f_k = \mu_k n$$

These three equations can be combined together as follows:

$$(1/4)Av^2 = mg\sin\theta - f_k = mg\sin\theta - \mu_k n = mg\sin\theta - \mu_k\, mg\cos\theta$$

$$\Rightarrow v_{\text{term}} = \left(mg\frac{\sin\theta - \mu_k\cos\theta}{\frac{1}{4}A}\right)^{1/2}$$

Using $\mu_k = 0.06$ and $A = 1.8\text{ m}\times 0.40\text{ m} = 0.72\text{ m}^2$, we find

$$v_{\text{term}} = \left[(80\text{ kg})(9.8\text{ m/s}^2)\left(\frac{\sin 40° - 0.06\cos 40°}{(\frac{1}{4}\text{ kg/m}^3)(0.72\text{ m}^2)}\right)\right]^{1/2} = 51\text{ m/s}$$

**Assess:** A terminal speed of 51 m/s corresponds to a speed of ≈100 mph. This speed is reasonable but high due to the steep slope angle of 40° and a small coefficient of friction.

**6.57. Model:** We will model the skier as a particle, and use the model of kinetic friction.
**Visualize:**

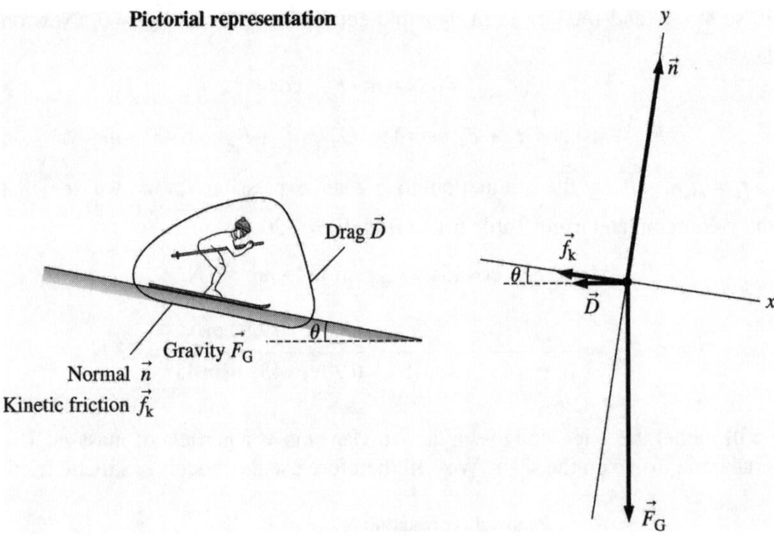


**Solve:** Your best strategy, if it's possible, is to travel at a very slow *constant* speed $(\vec{a} = \vec{0}$ so $\vec{F}_{\text{net}} = \vec{0})$. Alternatively, you want the smallest positive $a_x$. A negative $a_x$ would cause you to slow and stop. Let's find the value of $\mu_k$ that gives $\vec{F}_{\text{net}} = \vec{0}$. Newton's second law for the skier and the model of kinetic friction are

$$(F_{\text{net}})_x = \Sigma F_x = n_x + (F_G)_x + (f_k)_x + (D)_x = 0 + mg\sin\theta - f_k - D\cos\theta = 0\text{ N}$$

$$(F_{\text{net}})_y = \Sigma F_y = n_y + (F_G)_y + (f_k)_y + (D)_y = n - mg\cos\theta + 0\text{ N} - D\sin\theta = 0\text{ N}$$

$$f_k = \mu_k n$$

The *x*- and *y*-component equations are

$$f_k = +mg\sin\theta - D\cos\theta n = mg\cos\theta + D\sin\theta$$

From the model of kinetic friction,

$$\mu_k = \frac{f_k}{n} = \frac{mg\sin\theta - D\cos\theta}{mg\cos\theta + D\sin\theta} = \frac{75\text{ kg}\left(9.8\text{ m/s}^2\right)\sin 15° - \left(50\text{ N}\right)\cos 15°}{75\text{ kg}\left(9.8\text{ m/s}^2\right)\cos 15° + \left(50\text{ N}\right)\sin 15°} = 0.196$$

Yellow wax with $\mu_k = 0.20$ applied to skis will make the skis stick and hence cause the skier to stop. The skier's next choice is to use the green wax with $\mu_k = 0.15$.

**6.67. Solve:** **(a)** A 20.0 kg wooden crate is being pulled up a 20° wooden incline by a rope that is connected to an electric motor. The crate's acceleration is measured to be $2.0\text{ m/s}^2$. The coefficient of kinetic friction between the crate and the incline is 0.20. Find the tension $T$ in the rope.
**(b)**

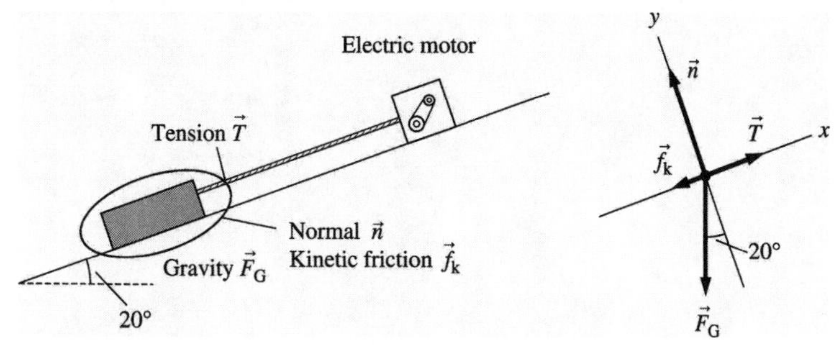


**(c)** Newton's second law for this problem in the component form is

$$(F_{net})_x = \Sigma F_x = T - 0.20n - (20\text{ kg})(9.80\text{ m/s}^2)\sin 20° = (20\text{ kg})(2.0\text{ m/s}^2)$$

$$(F_{net})_y = \Sigma F_y = n - (20\text{ kg})(9.80\text{ m/s}^2)\cos 20° = 0\text{ N}$$

Solving the $y$-component equation, $n = 184.18$ N. Substituting this value for $n$ in the $x$-component equation yields $T = 144$ N.

# NEWTON'S THIRD LAW

7

**7.1. Visualize:**

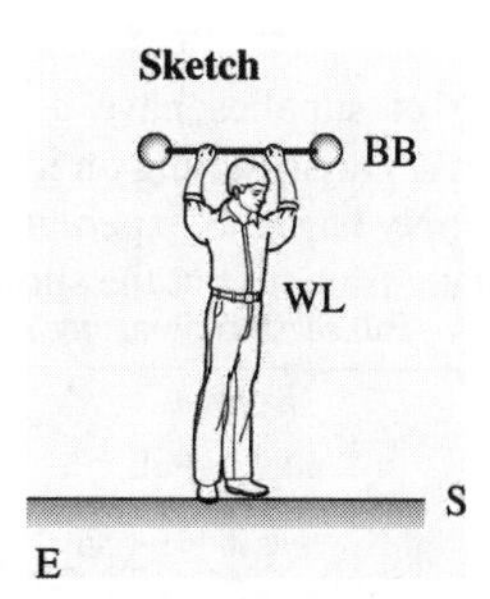


**Solve:** **(a)** The weight lifter is holding the barbell in dynamic equilibrium as he stands up, so the net force on the barbell and on the weight lifter must be zero. The barbells have an upward contact force from the weight lifter and the gravitational force downward. The weight lifter has a downward contact force from the barbells and an upward one from the surface. Gravity also acts on the weight lifter.

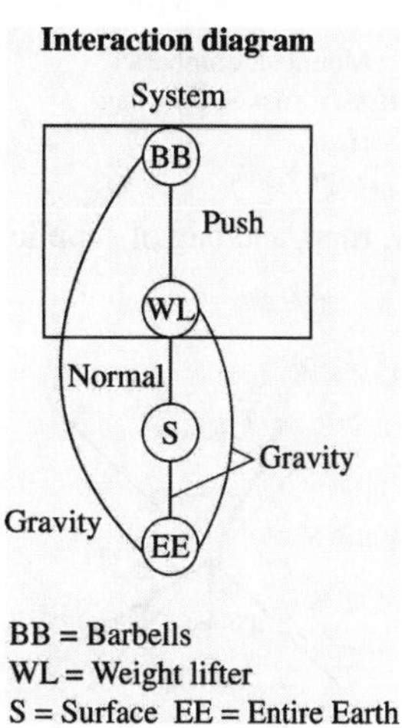


**(b)** The system is the weight lifter and barbell, as indicated in the figure.

**(c)**

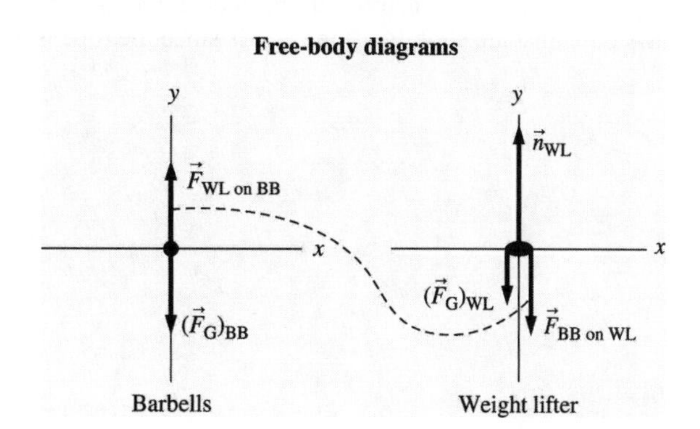

**7.3. Visualize:**

**Solve:** **(a)** Both the mountain climber and bag of supplies have a normal force from the surface on them, as well as a gravitational force vertically downward. The rope has gravity acting on it, along with pulls on each end from the mountain climber and supply bag. Both the mountain climber and supply bag also experience a frictional force with the surface of the mountain. In the case of the motionless mountain climber it is static friction, but the sliding supply bag experiences kinetic friction.

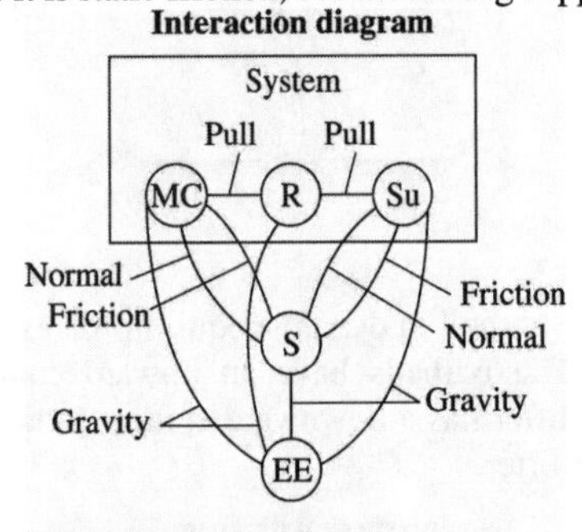


**(b)** The system consists of the mountain climber, rope, and bag of supplies, as indicated in the figure.

**(c)**

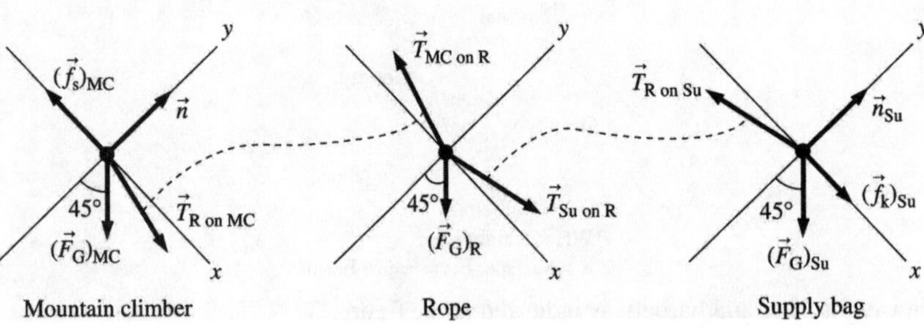


**Assess:** Since the motion is along the surface, it is convenient to choose the $x$-coordinate axis along the surface. The free-body diagram of the rope shows pulls that are slightly off the $x$-axis since the rope is not massless.

**7.5. Visualize:** Please refer to Figure EX7.5.

**Solve:** **(a)** Gravity acts on both blocks, and where Block A is in contact with the floor there is a normal force and friction. The string tension is the same on both blocks since the rope and pulley are massless, and the pulley is frictionless. There are two third law pairs of forces at the surface where the two blocks meet. Block B pushes against Block A with a normal force, while Block A has a reaction force that pushes back against Block B. There is also friction between the two blocks at the surface.

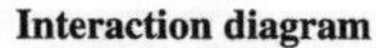


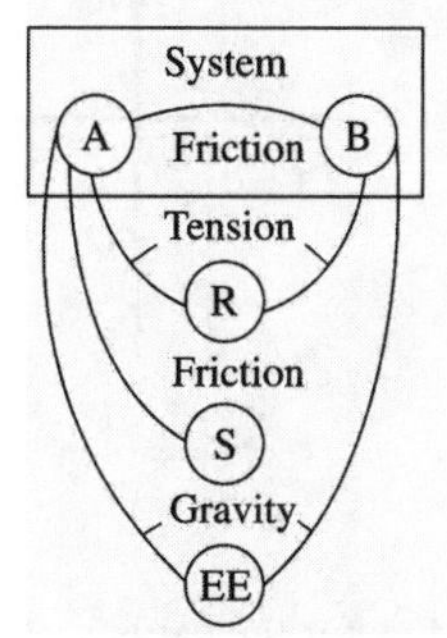


**(b)** A string that will not stretch constrains the two blocks to accelerate at the same rate but in opposite directions. Block A accelerates down the incline with the same acceleration that Block B has up the incline. The system consists of the two blocks, as indicated in the figure.

**(c)**

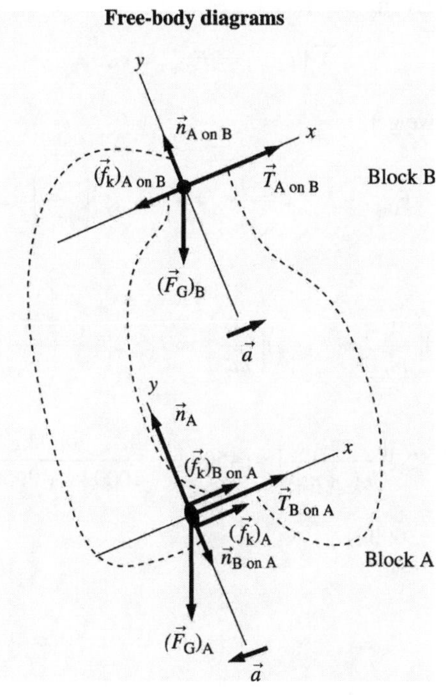


**Assess:** The inclined coordinate systems allow the acceleration $a$ to be purely along the $x$-axis. This is convenient since the one component of $a$ is zero, simplifying the mathematical expression of Newton's second law.

**7.9. Model:** The car and the truck will be modeled as particles and denoted by the symbols C and T, respectively. The surface of the ground will be denoted by the symbol S.
**Visualize:**

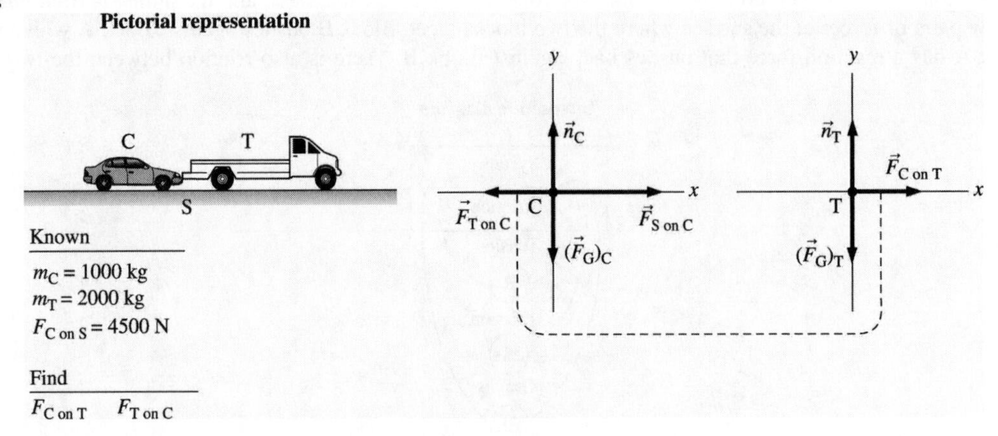


**Solve:** **(a)** The $x$-component of Newton's second law for the car is

$$\sum (F_{\text{on C}})_x = F_{\text{S on C}} - F_{\text{T on C}} = m_C a_C$$

The $x$-component of Newton's second law for the truck is

$$\sum (F_{\text{on T}})_x = F_{\text{C on T}} = m_T a_T$$

Using $a_C = a_T = a$ and $F_{\text{T on C}} = F_{\text{C on T}}$, we get

$$\left(F_{\text{C on S}} - F_{\text{C on T}}\right)\left(\frac{1}{m_C}\right) = a \quad \left(F_{\text{C on T}}\right)\left(\frac{1}{m_T}\right) = a$$

Combining these two equations,

$$\left(F_{\text{C on S}} - F_{\text{C on T}}\right)\left(\frac{1}{m_C}\right) = \left(F_{\text{C on T}}\right)\left(\frac{1}{m_T}\right) \Rightarrow F_{\text{C on T}}\left(\frac{1}{m_C} + \frac{1}{m_T}\right) = \left(F_{\text{C on S}}\right)\left(\frac{1}{m_C}\right)$$

$$\Rightarrow F_{\text{C on T}} = \left(F_{\text{C on S}}\right)\left(\frac{m_T}{m_C + m_T}\right) = (4500\text{ N})\left(\frac{2000\text{ kg}}{1000\text{ kg} + 2000\text{ kg}}\right) = 3000\text{ N}$$

**(b)** Due to Newton's third law, $F_{\text{T on C}} = 3000$ N.

**7.13. Model:** Together the carp (C) and the trout (T) make up the system that will be represented through the particle model. The fishing rod line (R) is assumed to be massless.

**Visualize:**

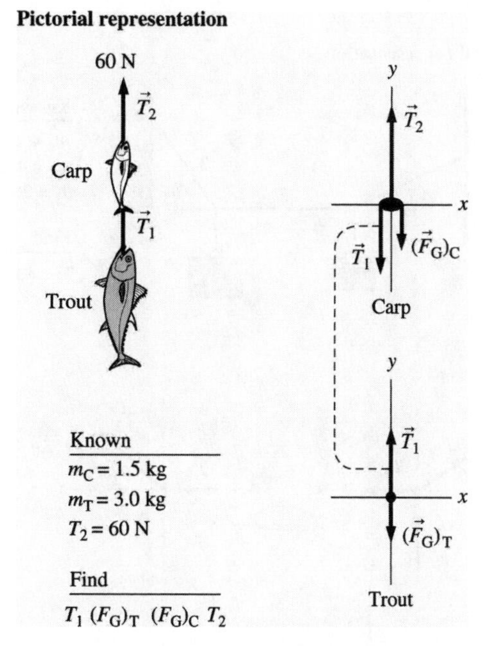


**Solve:** Jimmy's pull $T_2$ is larger than the total weight of the fish, so they accelerate upward. They are tied together, so each fish has the same acceleration $a$. Newton's second law along the $y$-direction for the carp and the trout is

$$\sum\left(F_{\text{on C}}\right)_y = T_2 - T_1 - (F_G)_C = m_C a \quad \sum\left(F_{\text{on T}}\right)_y = T_1 - (F_G)_T = m_T a$$

Adding these two equations gives

$$a = \frac{T_2 - (F_G)_C - (F_G)_T}{(m_C + m_T)} = \frac{60\text{ N} - (1.5\text{ kg})(9.8\text{ m/s}^2) - (3\text{ kg})(9.8\text{ m/s}^2)}{1.5\text{ kg} + 3.0\text{ kg}} = 3.533\text{ m/s}^2$$

Substituting this value of acceleration back into the force equation for the trout, we find that

$$T_1 = m_T(a+g) = (3\text{ kg})(3.533\text{ m/s}^2 + 9.8\text{ m/s}^2) = 40\text{ N}$$

$$(F_G)_T = m_T g = (3\text{ kg})(9.8\text{ m/s}^2) = 29.4\text{ N} \quad (F_G)_C = m_C g = (1.5\text{ kg})(9.8\text{ m/s}^2) = 14.7\text{ N}$$

Thus, $T_2 > T_1 > (F_G)_T > (F_G)_C$.

**7.17. Model:** The two hanging blocks, which can be modeled as particles, together with the two knots where rope 1 meets with rope 2 and rope 2 meets with rope 3 form a system. All the four objects in the system are in static equilibrium. The ropes are assumed to be massless.

**Visualize:**

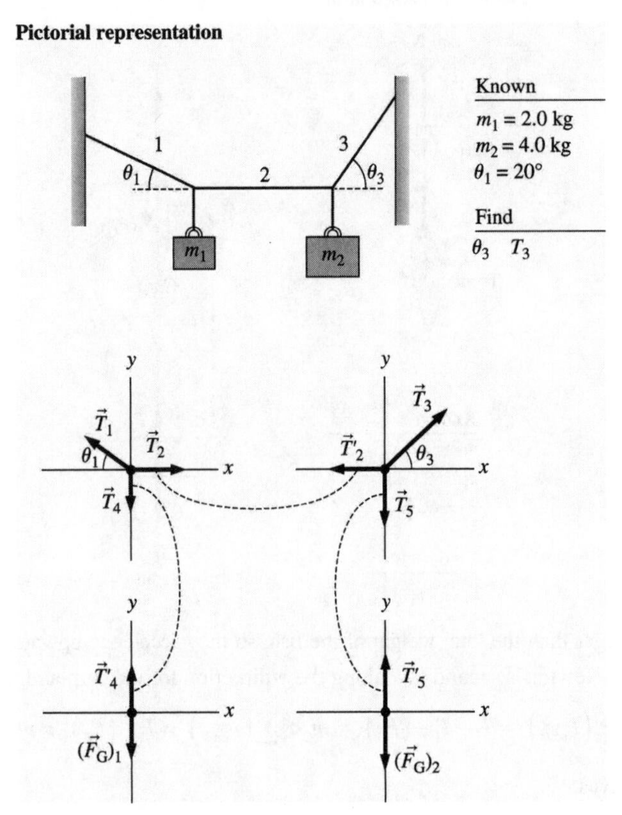


**Solve:** (a) We will consider both the two hanging blocks *and* the two knots. The blocks are in static equilibrium with $\vec{F}_{net} = 0$ N. Note that there are three action/reaction pairs. For Block 1 and Block 2, $\vec{F}_{net} = 0$ N and we have

$$T_4' = (F_G)_1 = m_1 g \quad T_5' = (F_G)_2 = m_2 g$$

Then, by Newton's third law:

$$T_4 = T_4' = m_1 g \quad T_5 = T_5' = m_2 g$$

The knots are also in equilibrium. Newton's law applied to the left knot is

$$(F_{net})_x = T_2 - T_1\cos\theta_1 = 0 \text{ N} \quad (F_{net})_y = T_1\sin\theta_1 - T_4 = T_1\sin\theta_1 - m_1 g = 0 \text{ N}$$

The $y$-equation gives $T_1 = m_1 g/\sin\theta_1$. Substitute this into the $x$-equation to find

$$T_2 = \frac{m_1 g\cos\theta_1}{\sin\theta_1} = \frac{m_1 g}{\tan\theta_1}$$

Newton's law applied to the right knot is

$$(F_{net})_x = T_3\cos\theta_3 - T_2' = 0 \text{ N} \quad (F_{net})_y = T_3\sin\theta_3 - T_5 = T_3\sin\theta_3 - m_2 g = 0 \text{ N}$$

These can be combined just like the equations for the left knot to give

$$T_2' = \frac{m_2 g\cos\theta_3}{\sin\theta_3} = \frac{m_2 g}{\tan\theta_3}$$

But the forces $\vec{T}_2$ and $\vec{T}_2'$ are an action/reaction pair, so $T_2 = T_2'$. Therefore,

$$\frac{m_1 g}{\tan\theta_1} = \frac{m_2 g}{\tan\theta_3} \Rightarrow \tan\theta_3 = \frac{m_2}{m_1}\tan\theta_1 \Rightarrow \theta_3 = \tan^{-1}(2\tan 20°) = 36°$$

We can now use the $y$-equation for the right knot to find $T_3 = m_2 g/\sin\theta_3 = 67$ N.

**7.23. Model:** Sled A, sled B, and the dog (D) are treated like particles in the model of kinetic friction.
**Visualize:**

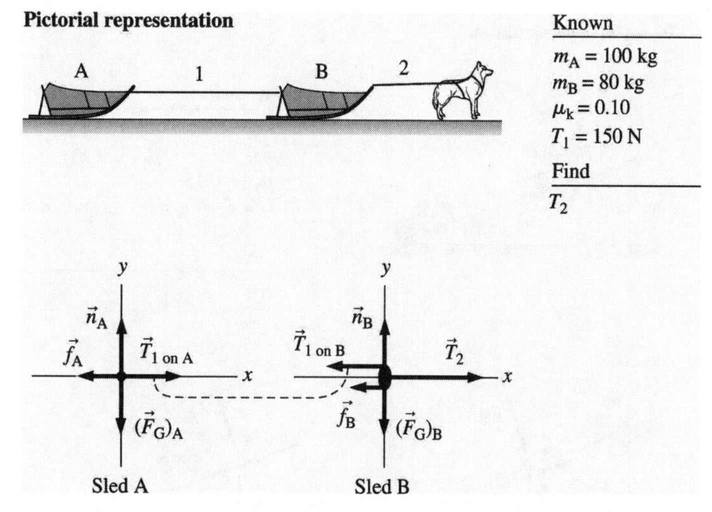


**Solve:** The acceleration constraint is $(a_A)_x = (a_B)_x = a_x$. Newton's second law on sled A is

$$\sum\left(\vec{F}_{\text{on A}}\right)_y = n_A - (F_G)_A = 0 \text{ N} \Rightarrow n_A = (F_G)_A = m_A g \quad \sum\left(\vec{F}_{\text{on A}}\right)_x = T_{1\text{ on A}} - f_A = m_A a_x$$

Using $f_A = \mu_k n_A$, the $x$-equation yields

$$T_{1\text{ on A}} - \mu_k n_A = m_A a_x \Rightarrow 150 \text{ N} - (0.1)(100 \text{ kg})(9.8 \text{ m/s}^2) = (100 \text{ kg})a_x \Rightarrow a_x = 0.52 \text{ m/s}^2$$

On sled B:

$$\sum\left(\vec{F}_{\text{on B}}\right)_y = n_B - (F_G)_B = 0 \text{ N} \Rightarrow n_B = (F_G)_B = m_B g \quad \sum\left(\vec{F}_{\text{on B}}\right)_x = T_2 - T_{1\text{ on B}} - f_B = m_B a_x$$

$T_{1\text{ on B}}$ and $T_{1\text{ on A}}$ act as if they are an action/reaction pair, so $T_{1\text{ on B}} = 150$ N. Using $f_B = \mu_k n_B = (0.10)(80 \text{ kg})(9.8 \text{ m/s}^2) = 78.4$ N, we get

$$T_2 - 150 \text{ N} - 78.4 \text{ N} = (80 \text{ kg})(0.52 \text{ m/s}^2) \Rightarrow T_2 = 270 \text{ N}$$

Thus the tension $T_2 = 2.7 \times 10^2$ N.

**7.29. Model:** Assume package A and package B are particles. Use the model of kinetic friction and the constant-acceleration kinematic equations.

**Visualize:**

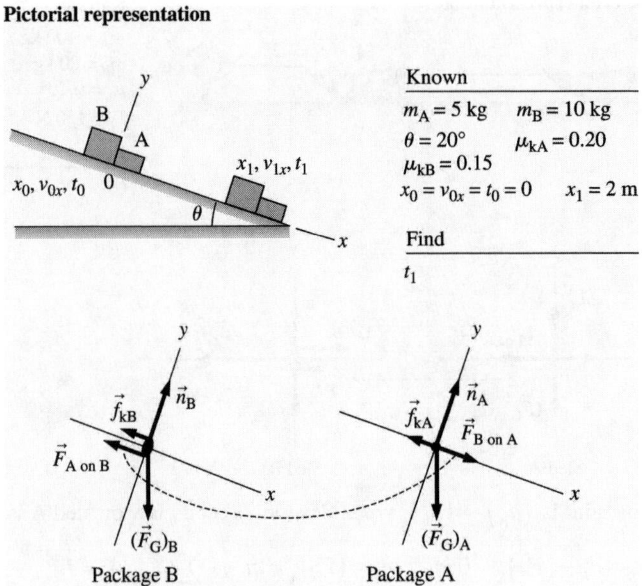


**Solve:** Package B has a smaller coefficient of friction. It will try to overtake package A and push against it. Package A will push back on B. The acceleration constraint is $(a_A)_x = (a_B)_x = a$.

Newton's second law for each package is

$$\sum(F_{\text{on A}})_x = F_{\text{B on A}} + (F_G)_A \sin\theta - f_{kA} = m_A a$$

$$\Rightarrow F_{\text{B on A}} + m_A g \sin\theta - \mu_{kA}(m_A g \cos\theta) = m_A a$$

$$\sum(F_{\text{on B}})_x = -F_{\text{A on B}} - f_{kB} + (F_G)_B \sin\theta = m_B a$$

$$\Rightarrow -F_{\text{A on B}} - \mu_{kB}(m_B g \cos\theta) + m_B g \sin\theta = m_B a$$

where we have used $n_A = m_A \cos\theta g$ and $n_B = m_B \cos\theta g$. Adding the two force equations, and using $F_{\text{A on B}} = F_{\text{B on A}}$ because they are an action/reaction pair, we get

$$a = g\sin\theta - \frac{(\mu_{kA} m_A + \mu_{kB} m_B)(g\cos\theta)}{m_A + m_B} = 1.82 \text{ m/s}^2$$

Finally, using $x_1 = x_0 + v_{0x}(t_1 - t_0) + \frac{1}{2}a(t_1 - t_0)^2$,

$$2.0 \text{ m} = 0 \text{ m} + 0 \text{ m} + \tfrac{1}{2}(1.82 \text{ m/s}^2)(t_1 - 0 \text{ s})^2 \Rightarrow t_1 = 1.48 \text{ s}$$

**7.35. Model:** The sled (S) and the box (B) will be treated in the particle model, and the model of friction will be used.
**Visualize:**

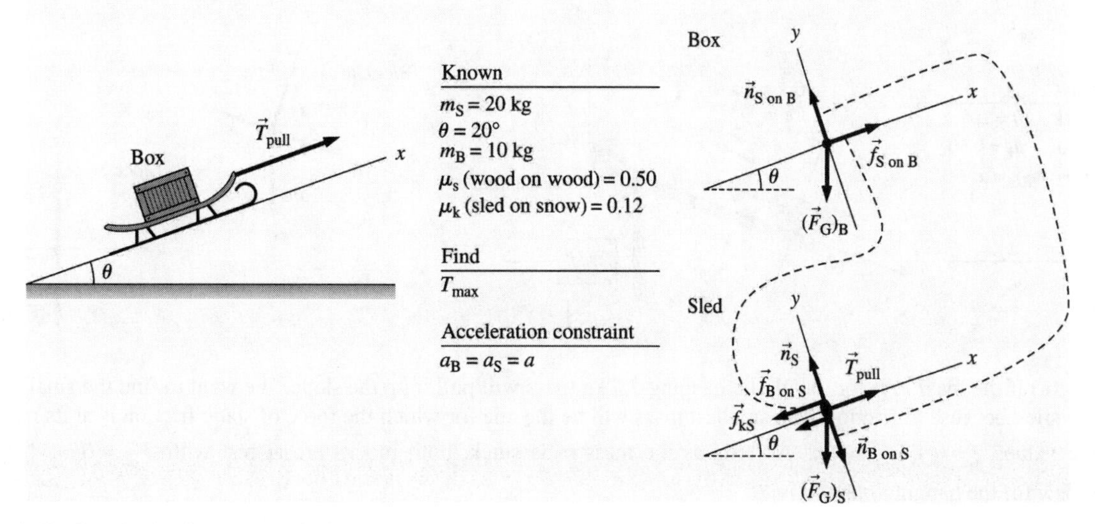


In the sled's free-body diagram $n_S$ is the normal (contact) force on the sled due to the snow. Similarly $f_{kS}$ is the force of kinetic friction on the sled due to snow.
**Solve:** Newton's second law on the box in the $y$-direction is

$$n_{S\ on\ B} - (F_G)_B \cos 20° = 0 \text{ N} \Rightarrow n_{S\ on\ B} = (10 \text{ kg})(9.8 \text{ m/s}^2)\cos 20° = 92.09 \text{ N}$$

The static friction force $\vec{f}_{S\ on\ B}$ accelerates the box. The maximum acceleration occurs when static friction reaches its maximum possible value.

$$(f_s)_{max} = \mu_s n_{S\ on\ B} = (0.50)(92.09 \text{ N}) = 46.05 \text{ N}$$

Newton's second law along the $x$-direction thus gives the maximum acceleration

$$f_{S\ on\ B} - (F_G)_B \sin 20° = m_B a \Rightarrow 46.05 \text{ N} - (10 \text{ kg})(9.8 \text{ m/s}^2)\sin 20° = (10 \text{ kg})a \Rightarrow a = 1.25 \text{ m/s}^2$$

Newton's second law for the sled along the $y$-direction is

$$n_S - n_{B\ on\ S} - (F_G)_S \cos 20° = 0 \text{ N}$$

$$\Rightarrow n_S = n_{B\ on\ S} + m_S g \cos 20° = (92.09 \text{ N}) + (20 \text{ kg})(9.8 \text{ m/s}^2)\cos 20° = 276.27 \text{ N}$$

Therefore, the force of friction on the sled by the snow is

$$f_{kS} = (\mu_k) n_S = (0.06)(276.27 \text{ N}) = 16.58 \text{ N}$$

Newton's second law along the $x$-direction is

$$T_{pull} - w_S \sin 20° - f_{kS} - f_{B\ on\ S} = m_S a$$

The friction force $f_{B\ on\ S} = f_{S\ on\ B}$ because these are an action/reaction pair. We're using the maximum acceleration, so the maximum tension is

$$T_{max} - (20 \text{ kg})(9.8 \text{ m/s}^2)\sin 20° - 16.58 \text{ N} - 46.05 \text{ N} = (20 \text{ kg})(1.25 \text{ m/s}^2)$$

$$\Rightarrow T_{max} = 155 \text{ N}$$

**7.39. Model:** Assume the particle model for the two blocks, and the model of kinetic and static friction.

**Visualize:**

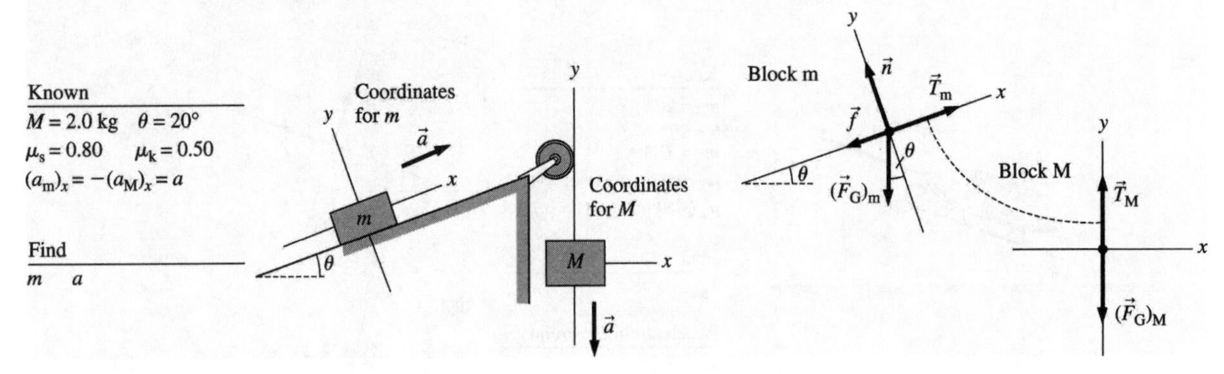


**Solve:** **(a)** If the mass $m$ is too small, the hanging 2.0 kg mass will pull it up the slope. We want to find the smallest mass that will stick because of friction. The smallest mass will be the one for which the force of static friction is at its maximum possible value: $f_s = (f_s)_{\max} = \mu_s n$. As long as the mass $m$ is stuck, both blocks are at rest with $\vec{F}_{\text{net}} = 0$ N. Newton's second law for the hanging mass $M$ is

$$(F_{\text{net}})_y = T_M - Mg = 0\text{ N} \Rightarrow T_M = Mg = 19.6\text{ N}$$

For the smaller mass $m$,

$$(F_{\text{net}})_x = T_m - f_s - mg\sin\theta = 0\text{ N} \quad (F_{\text{net}})_y = n - mg\cos\theta \Rightarrow n = mg\cos\theta$$

For a massless string and frictionless pulley, forces $\vec{T}_m$ and $\vec{T}_M$ act as if they are an action/reaction pair. Thus $T_m = T_M$. Mass $m$ is a minimum when $(f_s)_{\max} = \mu_s n = \mu_s mg\cos\theta$. Substituting these expressions into the $x$-equation,

$$T_M - \mu_s mg\cos\theta - mg\sin\theta = 0\text{ N} \Rightarrow m = \frac{T_M}{(\mu_s\cos\theta + \sin\theta)g} = 1.83\text{ kg}$$

**(b)** Because $\mu_k < \mu_s$ the 1.83 kg block will begin to slide up the ramp, and the 2.0 kg mass will begin to fall, if the block is nudged ever so slightly. Now the net force and the acceleration are *not* zero. Notice how, in the pictorial representation, we chose different coordinate systems for the two masses. This gives block M an acceleration with only a $y$-component and block m an acceleration with only an $x$-component. The magnitudes of the accelerations are the same because the blocks are tied together. But block M has a negative acceleration component $a_y$ (vector $\vec{a}$ points down) whereas block m has a positive $a_x$. Thus the acceleration constraint is $(a_m)_x = -(a_M)_y = a$, where $a$ will have a positive value. Newton's second law for block M is

$$(F_{\text{net}})_y = T - Mg = M(a_M)_y = -Ma$$

For block m we have

$$(F_{\text{net}})_x = T - f_k - mg\sin\theta = T - \mu_k mg\cos\theta - mg\sin\theta = m(a_m)_x = ma$$

In writing these equations, we used Newton's third law to write $T_m = T_M = T$. Also, we noticed that the $y$-equation and the friction model for block m don't change, except for $\mu_s$ becoming $\mu_k$, so we already know $f_k$ from part (a). Notice that the tension in the string is *not* the gravitational force $Mg$. We have two equations in the two unknowns $T$ and $a$:

$$T - Mg = -Ma \quad T - (\mu_k\cos\theta + \sin\theta)mg = ma$$

Subtracting the second equation from the first to eliminate $T$,

$$-Mg + (\mu_k\cos\theta + \sin\theta)mg = -Ma - ma = -(M + m)a$$

$$\Rightarrow a = \frac{M - (\mu_k\cos\theta + \sin\theta)m}{M + m}g = 1.32\text{ m/s}^2$$

**7.45. Model:** Use the particle model for the tightrope walker and the rope. The rope is assumed to be massless, so the tension in the rope is uniform.

**Visualize:**

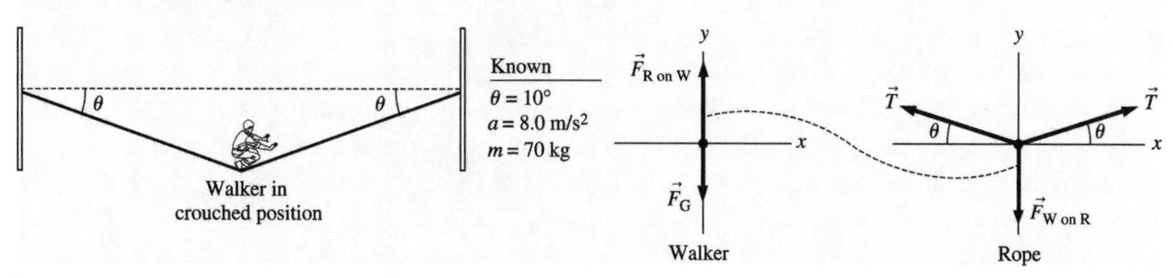


**Solve:** Newton's second law for the tightrope walker is

$$F_{\text{R on W}} - F_{\text{G}} = ma \Rightarrow F_{\text{R on W}} = m(a+g) = (70\text{ kg})(8.0\text{ m/s}^2 + 9.8\text{ m/s}^2) = 1.25\times10^3\text{ N}$$

Now, Newton's second law for the rope gives

$$\sum(F_{\text{on R}})_y = T\sin\theta + T\sin\theta - F_{\text{W on R}} = 0\text{ N} \Rightarrow T = \frac{F_{\text{W on R}}}{2\sin10°} = \frac{F_{\text{R on W}}}{2\sin10°} = \frac{1.25\times10^3\text{ N}}{2\sin10°} = 3.6\times10^3\text{ N}$$

We used $F_{\text{W on R}} = F_{\text{R on W}}$ because they are an action/reaction pair.

# 8 DYNAMICS II: MOTION IN A PLANE

**8.1. Model:** The model rocket and the target will be treated as particles. The kinematics equations in two dimensions apply.
**Visualize:**

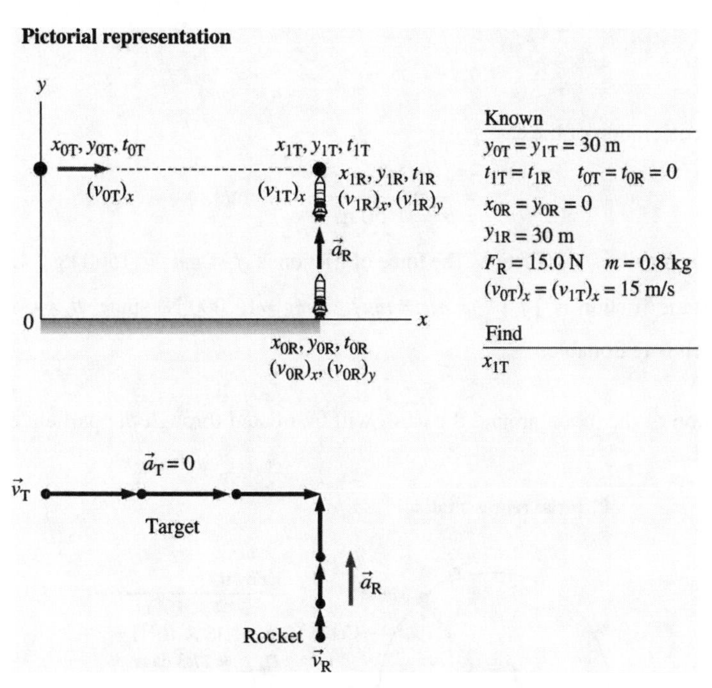


**Solve:** For the rocket, Newton's second law along the *y*-direction is

$$(F_{\text{net}})_y = F_R - mg = ma_R$$

$$\Rightarrow a_R = \frac{1}{m}(F_R - mg) = \frac{1}{0.8 \text{ kg}}\left[15 \text{ N} - (0.8 \text{ kg})(9.8 \text{ m/s}^2)\right] = 8.95 \text{ m/s}^2$$

Using the kinematic equation $y_{1R} = y_{0R} + (v_{0R})_y (t_{1R} - t_{0R}) + \frac{1}{2}a_R (t_{1R} - t_{0R})^2$,

$$30 \text{ m} = 0 \text{ m} + 0 \text{ m} + \tfrac{1}{2}(8.95 \text{ m/s}^2)(t_{1R} - 0 \text{ s})^2 \Rightarrow t_{1R} = 2.589 \text{ s}$$

For the target (noting $t_{1T} = t_{1R}$),

$$x_{1T} = x_{0T} + (v_{0T})_x (t_{1T} - t_{0T}) + \tfrac{1}{2}a_T (t_{1T} - t_{0T})^2 = 0 \text{ m} + (15 \text{ m/s})(2.589 \text{ s} - 0 \text{ s}) + 0 \text{ m} = 39 \text{ m}$$

You should launch when the target is 39 m away.
**Assess:** The rocket is to be fired when the target is at $x_{0T}$. For a net acceleration of approximately 9 m/s$^2$ in the vertical direction and a time of 2.6 s to cover a vertical distance of 30 m, a horizontal distance of 39 m is reasonable.

**8.5. Model:** We will use the particle model for the car which is in uniform circular motion.
**Visualize:**

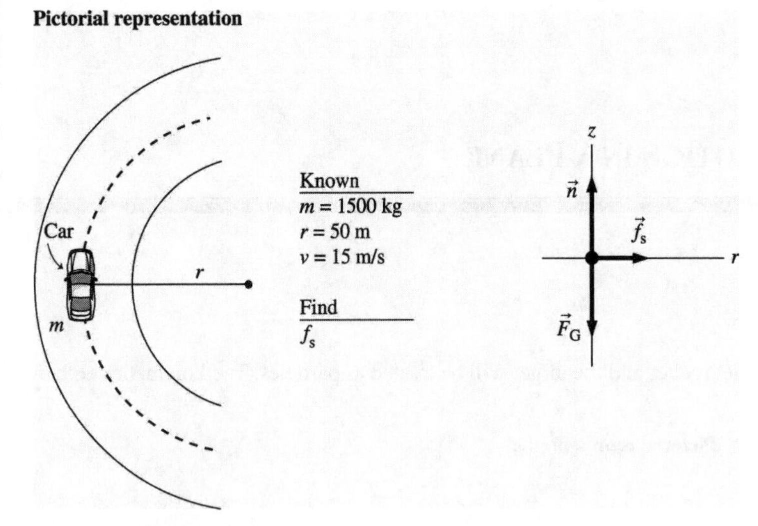


**Solve:** The centripetal acceleration of the car is

$$a_r = \frac{v^2}{r} = \frac{(15\text{ m/s})^2}{50\text{ m}} = 4.5\text{ m/s}^2$$

The acceleration is due to the force of static friction. The force of friction is $f_s = ma_r = (1500\text{ kg})(4.5\text{m/s}^2) = 6750\text{ N} =$ 6.8 kN.

**Assess:** The model of static friction is $(f_s)_{max} = n\mu_s = mg\mu_s \approx mg \approx 15{,}000\text{ N}$ since $\mu_s \approx 1$ for a dry road surface. We see that $f_s < (f_s)_{max}$, which is reasonable.

**8.9. Model:** The motion of the moon around the earth will be treated through the particle model. The circular motion is uniform.
**Visualize:**

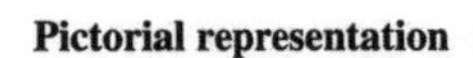


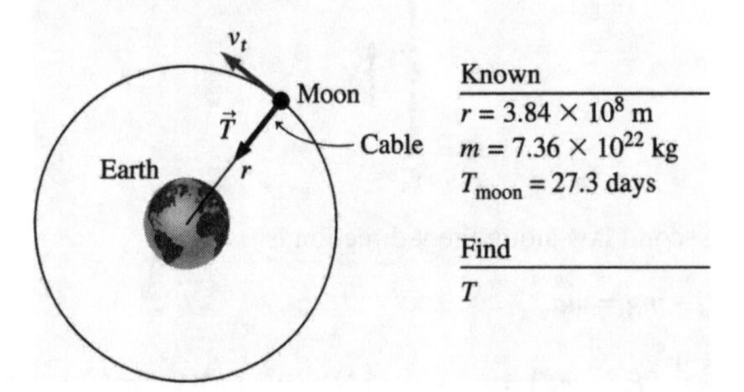


**Solve:** The tension in the cable provides the centripetal acceleration. Newton's second law is

$$\sum F_r = T = mr\omega^2 = mr\left(\frac{2\pi}{T_{\text{moon}}}\right)^2$$

$$= (7.36\times10^{22}\text{ kg})(3.84\times10^8\text{ m})\left[\frac{2\pi}{27.3\text{ days}}\times\frac{1\text{ day}}{24\text{ h}}\times\frac{1\text{ h}}{3600\text{ s}}\right]^2 = 2.01\times10^{20}\text{ N}$$

**Assess:** This is a tremendous tension, but clearly understandable in view of the moon's large mass and the large radius of circular motion around the earth.

**8.11. Model:** The satellite is considered to be a particle in uniform circular motion around the moon.
**Visualize:**

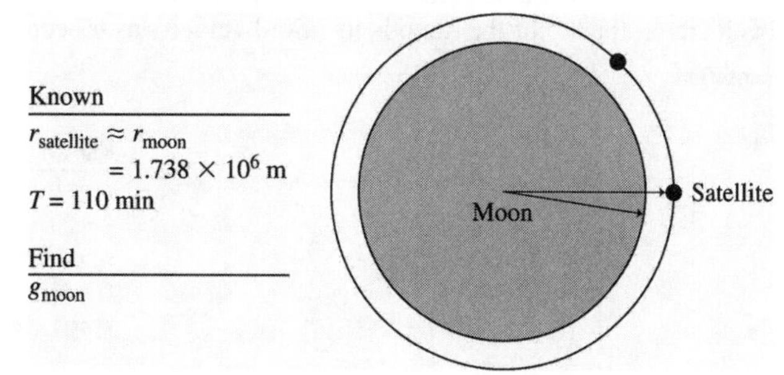


**Solve:** The radius of the moon is $1.738\times10^6$ m and the satellite's distance from the center of the moon is the same quantity. The angular velocity of the satellite is

$$\omega = \frac{2\pi}{T} = \frac{2\pi \text{ rad}}{110 \text{ min}} \times \frac{1 \text{min}}{60 \text{ s}} = 9.52\times10^{-4} \text{ rad/s}$$

and the centripetal acceleration is

$$a_r = r\omega^2 = (1.738\times10^6 \text{ m})(9.52\times10^{-4} \text{ rad/s})^2 = 1.58 \text{ m/s}^2$$

The acceleration of a body in orbit is the local "$g$" experienced by that body.

**8.17. Model:** Model the bucket of water as a particle in uniform circular motion.
**Visualize:**

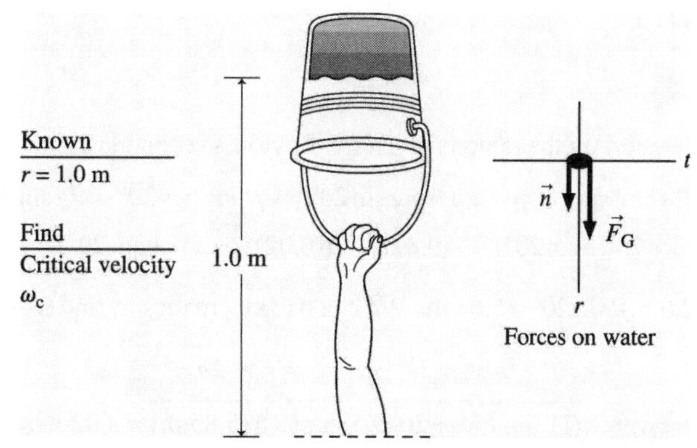


**Solve:** Let us say the distance from the bucket handle to the top of the water in the bucket is 35 cm. This makes the shoulder to water distance $65 \text{ cm} + 35 \text{ cm} = 1.00$ m. The minimum angular velocity for swinging a bucket of water in a vertical circle without spilling any water corresponds to the case when the speed of the bucket is critical. In this case, $n = 0$ N when the bucket is in the top position of the circular motion. We get

$$\sum F_r = n + F_G = 0 \text{ N} + mg = \frac{mv_c^2}{r} = mr\omega_c^2$$

$$\Rightarrow \omega_c = \sqrt{g/r} = \sqrt{\frac{9.8 \text{ m/s}^2}{1.00 \text{ m}}} = 3.13 \text{ rad/s} = 3.13 \text{ rad/s} \times \frac{1 \text{ rev}}{2\pi \text{ rad}} \times \frac{60 \text{ s}}{1 \text{ min}} = 30 \text{ rpm}$$

**8.23. Model:** Treat the motorcycle and rider as a particle.
**Visualize:** This is a two-part problem. Use an *s*-axis parallel to the slope for the first part, regular *xy*-coordinates for the second. The motorcycle's final velocity at the top of the ramp is its initial velocity as it becomes airborne.

**Pictorial representation**

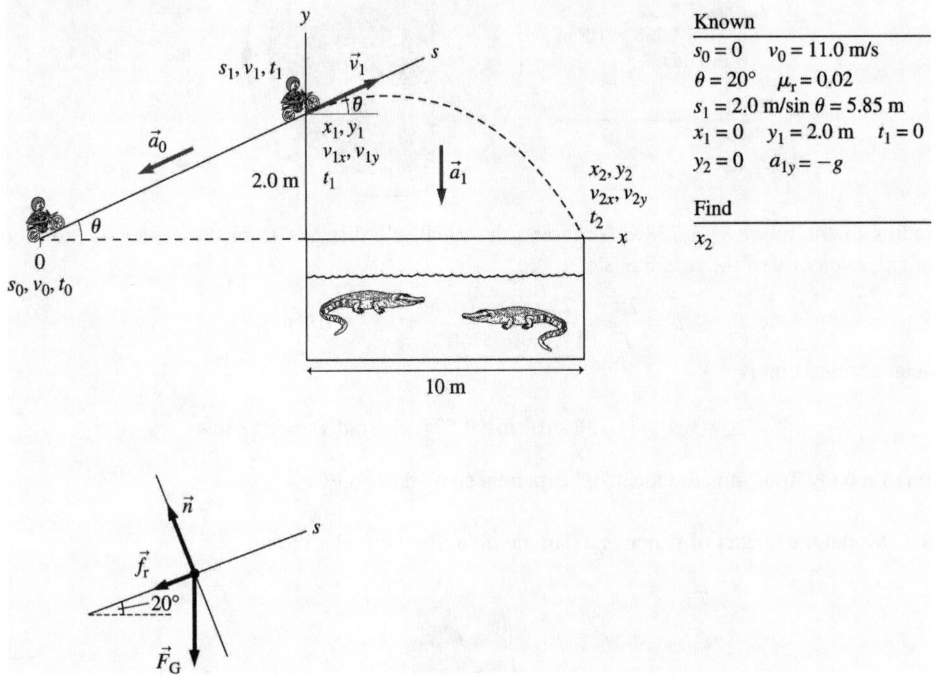

Known
$s_0 = 0$ $v_0 = 11.0$ m/s
$\theta = 20°$ $\mu_r = 0.02$
$s_1 = 2.0$ m/sin $\theta = 5.85$ m
$x_1 = 0$ $y_1 = 2.0$ m $t_1 = 0$
$y_2 = 0$ $a_{1y} = -g$
Find
$x_2$
$s_1, v_1, t_1$
$x_1, y_1$
$v_{1x}, v_{1y}$
$t_1$
$x_2, y_2$
$v_{2x}, v_{2y}$
$t_2$
2.0 m
10 m
0
$s_0, v_0, t_0$
20°


**Solve:** The motorcycle's acceleration on the ramp is given by Newton's second law:

$$(F_{net})_s = -f_r - mg\sin 20° = -\mu_r n - mg\sin 20° = -\mu_r mg\cos 20° - mg\sin 20° = ma_0$$
$$a_0 = -g(\mu_r\cos 20° + \sin 20°) = -(9.8\text{ m/s}^2)((0.02)\cos 20° + \sin 20°) = -3.536\text{ m/s}^2$$

The length of the ramp is $s_1 = (2.0\text{ m})/\sin 20° = 5.85$ m. We can use kinematics to find its speed at the top of the ramp:

$$v_1^2 = v_0^2 + 2a_0(s_1 - s_0) = v_0^2 + 2a_0 s_1$$
$$\Rightarrow v_1 = \sqrt{(11.0\text{ m/s})^2 + 2(-3.536\text{ m/s}^2)(5.85\text{ m})} = 8.92\text{ m/s}$$

This is the motorcycle's initial speed into the air, with velocity components $v_{1x} = v_1\cos 20° = 8.38$ m/s and $v_{1y} = v_1\sin 20° =$ 3.05 m/s. We can use the *y*-equation of projectile motion to find the time in the air:

$$y_2 = 0\text{ m} = y_1 + v_{1y}t_2 + \tfrac{1}{2}a_{1y}t_2^2 = 2.0\text{ m} + (3.05\text{ m/s})t_2 - (4.90\text{ m/s}^2)t_2^2$$

This quadratic equation has roots $t_2 = -0.399$ s (unphysical) and $t_2 = 1.021$ s. The *x*-equation of motion is thus

$$x_2 = x_1 + v_{1x}t_2 = 0\text{ m} + (8.38\text{ m/s})t_2 = 8.56\text{ m}$$

$8.56\text{ m} < 10.0\text{ m}$, so it looks like crocodile food.

**8.29. Model:** Use the particle model for the (cart + child) system which is in uniform circular motion.

**Visualize:**

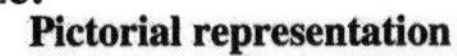


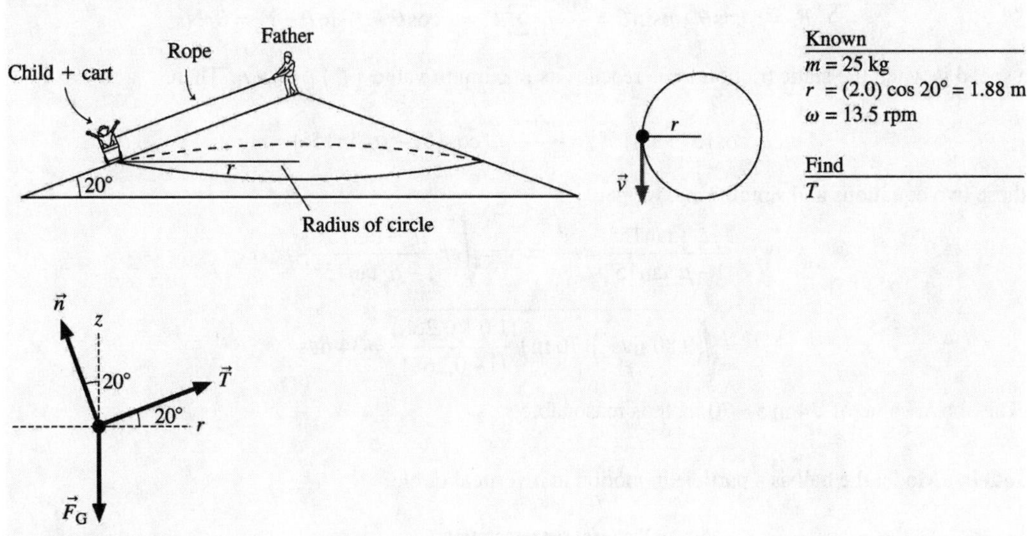


**Solve:** Newton's second law along $r$ and $z$ directions can be written:

$$\sum F_r = T\cos 20° - n\sin 20° = ma_r \quad \sum F_z = T\sin 20° - n\cos 20° - mg = 0$$

The cart's centripetal acceleration is

$$a_r = r\omega^2 = (2.0\cos 20° \text{ m})\left(13.5\ \frac{\text{rev}}{\text{min}} \times \frac{1\text{ min}}{60\text{ s}} \times \frac{2\pi\text{ rad}}{1\text{ rev}}\right)^2 = 3.756\text{ m/s}^2$$

The above force equations can be rewritten as

$$0.94T - 0.342n = (25\text{ kg})(3.756\text{ m/s}^2) = 93.9\text{ N}$$

$$0.342T + 0.94n = (25\text{ kg})(9.8\text{ m/s}^2) = 245\text{ N}$$

Solving these two equations yields $T = 172$ N for the tension in the rope.

**Assess:** In view of the child + cart weight of 245 N, a tension of 172 N is reasonable.

**8.31. Model:** We will use the particle model for the car, which is undergoing uniform circular motion on a banked highway, and the model of static friction.

**Visualize:**

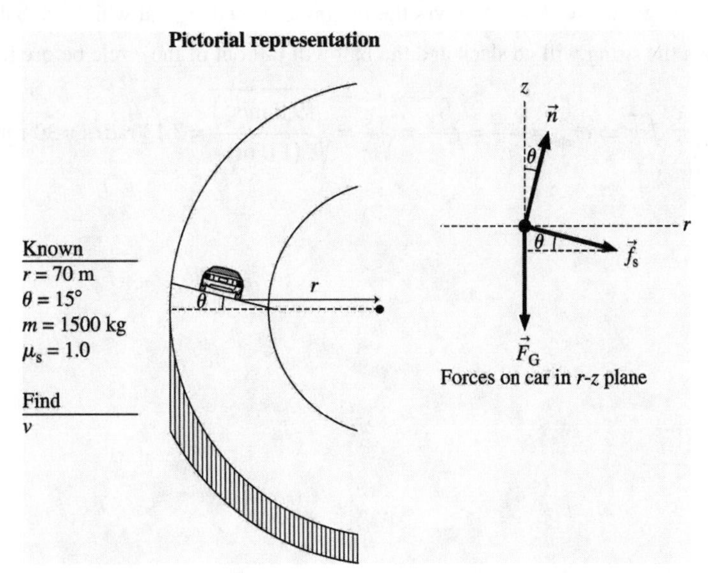

Note that we need to use the coefficient of static friction $\mu_s$, which is 1.0 for rubber on concrete.

**Solve:** Newton's second law for the car is

$$\sum F_r = f_s\cos\theta + n\sin\theta = \frac{mv^2}{r} \quad \sum F_z = n\cos\theta - f_s\sin\theta - F_G = 0 \text{ N}$$

Maximum speed is when the static friction force reaches its maximum value $(f_s)_{max} = \mu_s n$. Then

$$n(\mu_s\cos15° + \sin15°) = \frac{mv^2}{r} \quad n(\cos15° - \mu_s\sin15°) = mg$$

Dividing these two equations and simplifying, we get

$$\frac{\mu_s + \tan15°}{1-\mu_s\tan15°} = \frac{v^2}{gr} \Rightarrow v = \sqrt{gr\frac{\mu_s+\tan15°}{1-\mu_s\tan15°}}$$

$$= \sqrt{(9.80 \text{ m/s}^2)(70 \text{ m})\frac{(1.0+0.268)}{(1-0.268)}} = 34 \text{ m/s}$$

**Assess:** The above value of $34 \text{ m/s} \approx 70 \text{ mph}$ is reasonable.

**8.43. Model:** Model the ball as a particle in motion in a vertical circle.
**Visualize:**

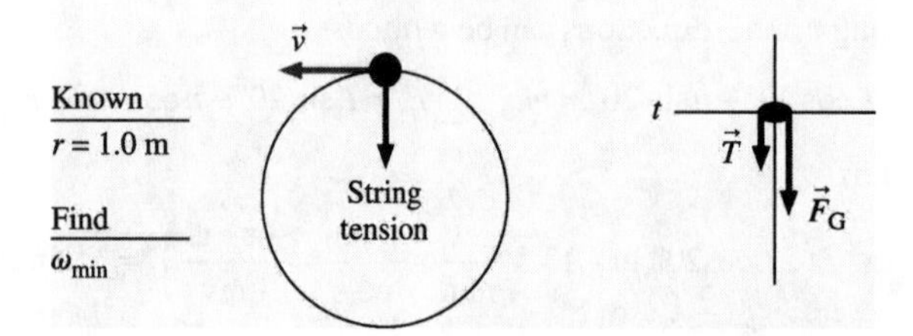


**Solve:** *If* the ball moves in a complete circle, then there is a tension force $\vec{T}$ when the ball is at the top of the circle. The tension force adds to the gravitational force to cause the centripetal acceleration. The forces are along the $r$-axis, and the center of the circle is below the ball. Newton's second law at the top is

$$(F_{net})_r = T + F_G = T + mg = \frac{mv^2}{r}$$

$$\Rightarrow v_{top} = \sqrt{rg + \frac{rT}{m}}$$

The tension $T$ can't become negative, so $T = 0$ N gives the minimum speed $v_{min}$ at which the ball moves in a circle. If the speed is less than $v_{min}$, then the string will go slack and the ball will fall out of the circle before it reaches the top. Thus,

$$v_{min} = \sqrt{rg} \Rightarrow \omega_{min} = \frac{v_{min}}{r} = \frac{\sqrt{rg}}{r} = \sqrt{\frac{g}{r}} = \sqrt{\frac{(9.8 \text{ m/s}^2)}{(1.0 \text{ m})}} = 3.13 \text{ rad/s} = 30 \text{ rpm}$$

**8.47. Model:** Model the ball as a particle swinging in a vertical circle, then as a projectile.
**Visualize:**

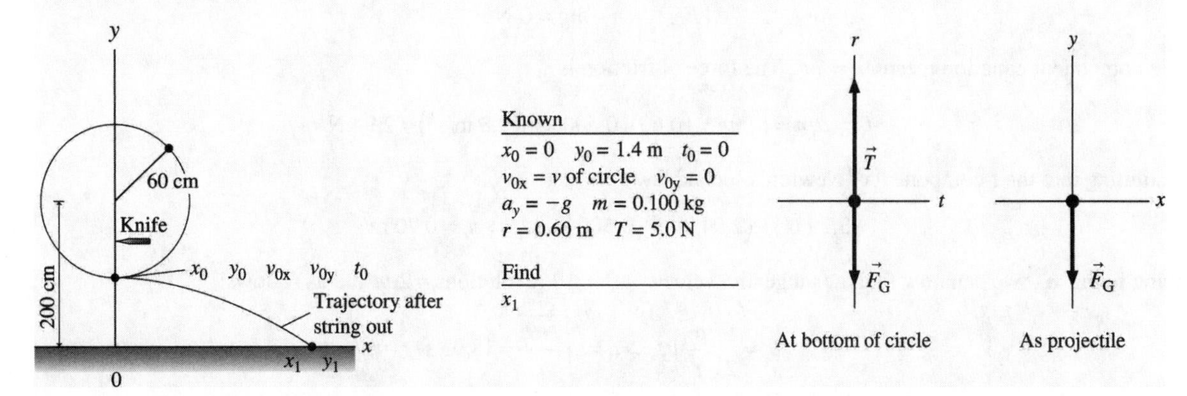


**Solve:** Initially, the ball is moving in a circle. Once the string is cut, it becomes a projectile. The final circular-motion velocity is the initial velocity for the projectile. The free-body diagram for circular motion is shown at the bottom of the circle. Since $T > F_G$, there is a net force toward the center of the circle that causes the centripetal acceleration. The $r$-equation of Newton's second law is

$$(F_{\text{net}})_r = T - F_G = T - mg = \frac{mv^2}{r}$$

$$\Rightarrow v_{\text{bottom}} = \sqrt{\frac{r}{m}(T - mg)} = \sqrt{\frac{0.60 \text{ m}}{0.100 \text{ kg}}\left[5.0 \text{ N} - (0.10 \text{ kg})(9.8 \text{ m/s}^2)\right]} = 4.91 \text{ m/s}$$

As a projectile the ball starts at $y_0 = 1.4$ m with $\vec{v}_0 = 4.91\hat{i}$ m/s. The equation for the $y$-motion is

$$y_1 = 0 \text{ m} = y_0 + v_{0y}\Delta t - \tfrac{1}{2}g(\Delta t)^2 = y_0 - \tfrac{1}{2}gt_1^2$$

This is easily solved to find that the ball hits the ground at time

$$t_1 = \sqrt{\frac{2y_0}{g}} = 0.535 \text{ s}$$

During this time interval it travels a horizontal distance

$$x_1 = x_0 + v_{0x}t_1 = (4.91 \text{ m/s})(0.535 \text{ s}) = 2.63 \text{ m}$$

So the ball hits the floor 2.6 m to the right of the point where the string was cut.

**8.51. Model:** Model the steel block as a particle and use the model of kinetic friction.
**Visualize:**

Pictorial representation

Known
$m = 500$ g
$r = 2.0$ m
$F = 3.5$ N
$\theta = 20°$ $\mu_k = 0.60$
$t_0 = 0$ $v_{0t} = 0$ $\theta_0 = 0$
$\theta_1 = 10$ rev

Find
$\omega_1$ $T_1$

$\vec{v}_t$ $\theta$ Steel block 70° Air Rod $r$ Pivot

t $\theta$ $\vec{F}$ $\vec{f}_k$ $\vec{T}$ r

Top view

z $\vec{n}$ $\vec{T}$ r $\vec{F}_G$

Edge view

**Solve:** **(a)** The components of thrust $(\vec{F})$ along the $r$-, $t$-, and $z$-directions are

$$F_r = F\sin 20° = (3.5 \text{ N})\sin 20° = 1.20 \text{ N} \quad F_t = F\cos 20° = (3.5 \text{ N})\cos 20° = 3.29 \text{ N} \quad F_z = 0 \text{ N}$$

Newton's second law is

$$(F_{\text{net}})_r = T + F_r = mr\omega^2 \quad (F_{\text{net}})_t = F_t - f_k = ma_t$$

$$(F_{\text{net}})_z = n - mg = 0 \text{ N}$$

The $z$-component equation means $n = mg$. The force of friction is

$$f_k = \mu_k n = \mu_k mg = (0.60)(0.500 \text{ kg})(9.8 \text{ m/s}^2) = 2.94 \text{ N}$$

Substituting into the $t$-component of Newton's second law

$$(3.29 \text{ N}) - (2.94 \text{ N}) = (0.500 \text{ kg})a_t \Rightarrow a_t = 0.70 \text{ m/s}^2$$

Having found $a_t$, we can now find the tangential velocity after 10 revolutions $= 20\pi$ rad as follows:

$$\theta_1 = \frac{1}{2}\left(\frac{a_t}{r}\right)t_1^2 \Rightarrow t_1 = \sqrt{\frac{2r\theta_1}{a_t}} = 18.95 \text{ s}$$

$$\omega_1 = \omega_0 + \left(\frac{a_t}{r}\right)t_1 = 6.63 \text{ rad/s}$$

The block's angular velocity after 10 s is 6.6 rad/s.
**(b)** Substituting $\omega_1$ into the $r$-component of Newton's second law yields:

$$T_1 + F_r = mr\omega_1^2 \Rightarrow T_1 + (1.20 \text{ N}) = (0.500 \text{ kg})(2.0 \text{ m})(6.63 \text{ rad/s})^2 \Rightarrow T_1 = 43 \text{ N}$$

9

# IMPULSE AND MOMENTUM

**9.5. Visualize:** Please refer to Figure EX9.5.
**Solve:** The impulse is defined in Equation 9.6 as

$$J_x = \int_{t_i}^{t_f} F_x(t)dt = \text{ area under the } F_x(t) \text{ curve between } t_i \text{ and } t_f$$

$$\Rightarrow 6.0 \text{ N s} = \tfrac{1}{2}(F_{max})(8 \text{ ms}) \Rightarrow F_{max} = 1.5\times10^3 \text{ N}$$

**9.7. Model:** Model the object as a particle and the interaction with the force as a collision.
**Visualize:** Please refer to Figure EX9.7.
**Solve:** Using the equations

$$p_{fx} = p_{ix} + J_x \text{ and } J_x = \int_{t_i}^{t_f} F_x(t)\,dt = \text{area under force curve}$$

$$(2.0 \text{ kg})v_{fx} = (2.0 \text{ kg})(1.0 \text{ m/s}) + \text{(area under the force curve)}$$

$$\Rightarrow v_{fx} = (1.0 \text{ m/s}) + \frac{1}{2.0 \text{ kg}}(1.0 \text{ s})(2.0 \text{ N}) = 2.0 \text{ m/s}$$

**Assess:** For an object with positive velocity, a positive impulse increases the object's speed. The opposite is true for an object with negative velocity.

**9.9. Model:** Use the particle model for the sled, the model of kinetic friction, and the impulse-momentum theorem.
**Visualize:**

Pictorial representation

$v_{ix} = 8.0$ m/s
$\vec{f}_k$
$v_{fx} = 5.0$ m/s
Find
$\Delta t = t_f - t_i$
$\mu_k = 0.25$
Before
After
$\vec{n}$
$\vec{f}_k$
$\vec{F}_G$

Note that the force of kinetic friction $f_k$ imparts a negative impulse to the sled.
**Solve:** Using $\Delta p_x = J_x$, we have

$$p_{fx} - p_{ix} = \int_{t_i}^{t_f} F_x(t)\,dt = -f_k \int_{t_i}^{t_f} dt = -f_k\Delta t \Rightarrow mv_{fx} - mv_{ix} = -\mu_k n\Delta t = -\mu_k mg\Delta t$$

We have used the model of kinetic friction $f_k = \mu_k n$, where $\mu_k$ is the coefficient of kinetic friction and $n$ is the normal (contact) force by the surface. The force of kinetic friction is independent of time and was therefore taken out of the impulse integral. Thus,

$$\Delta t = \frac{1}{\mu_k g}(v_{ix} - v_{fx}) = \frac{1}{(0.25)(9.8 \text{ m/s}^2)}(8.0 \text{ m/s} - 5.0 \text{ m/s}) = 1.22 \text{ s}$$

**9.15. Model:** Choose car + rainwater to be the system.
**Visualize:**

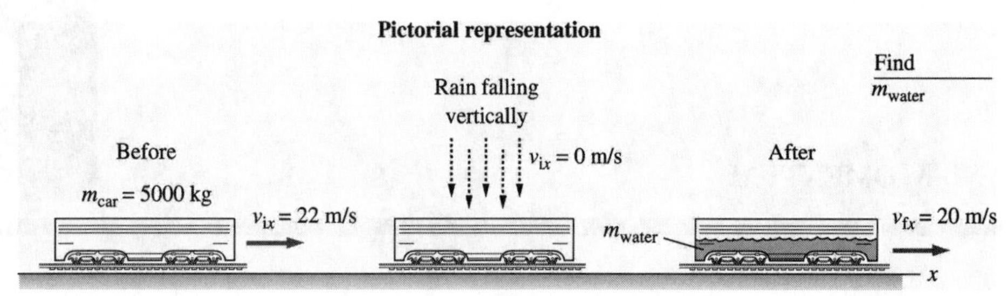


There are no *external* horizontal forces on the car + water system, so the horizontal momentum is conserved.
**Solve:** Conservation of momentum is $p_{fx} = p_{ix}$. Hence,

$$(m_{car} + m_{water})(20 \text{ m/s}) = (m_{car})(22 \text{ m/s}) + (m_{water})(0 \text{ m/s})$$

$$\Rightarrow (5000 \text{ kg} + m_{water})(20 \text{ m/s}) = (5000 \text{ kg})(22 \text{ m/s}) \Rightarrow m_{water} = 5.0 \times 10^2 \text{ kg}$$

**9.19. Model:** Because of external friction and drag forces, the car and the blob of sticky clay are not exactly an isolated system. But during the collision, friction and drag are not going to be significant. The momentum of the system will be conserved in the collision, within the impulse approximation.
**Visualize:**

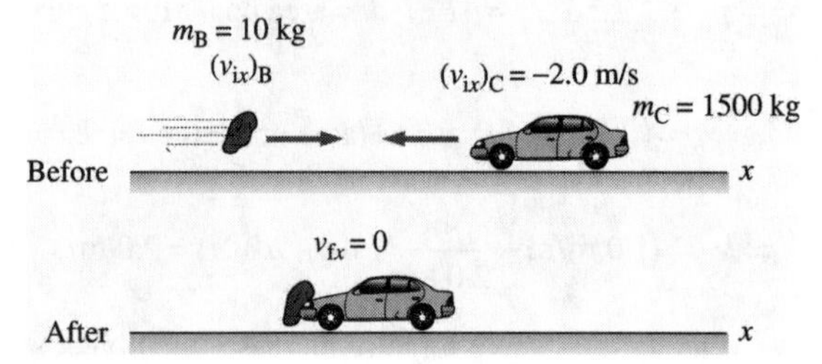


**Solve:** The conservation of momentum equation $p_{fx} = p_{ix}$ is

$$(m_C + m_B)(v_f)_x = m_B (v_{ix})_B + m_C (v_{ix})_C$$

$$\Rightarrow 0 \text{ kg m/s} = (10 \text{ kg})(v_{ix})_B + (1500 \text{ kg})(-2.0 \text{ m/s}) \Rightarrow (v_{ix})_B = 3.0 \times 10^2 \text{ m/s}$$

**Assess:** This speed of the blob is around 600 mph, which is very large. However, we must point out that a very large speed is *expected* in order to stop a car with only 10 kg of clay.

**9.21. Model:** We will define our system to be Bob + rock. Bob's (B) force on the rock (R) is equal to the rock's force on Bob. These are internal forces within the system. Bob is standing on frictionless ice, and the normal force by ice on the system balances the weight. $\vec{F}_{ext} = \vec{0}$ on the system, and thus momentum is conserved.
**Visualize:**

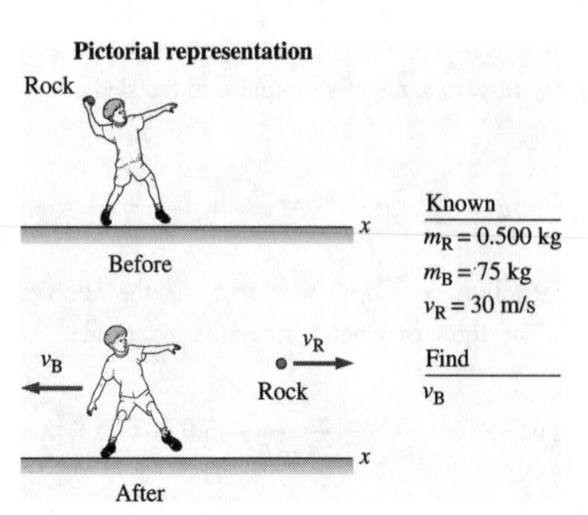

The initial momentum $p_{ix}$ of the system is zero because Bob and the rock are at rest. Thus $p_{fx} = 0$ kg m/s.

**Solve:** We have $m_B v_B + m_R v_R = 0$ kg m/s. Hence,

$$v_B = -\frac{m_R}{m_B} v_R = -\left(\frac{0.500 \text{ kg}}{75 \text{ kg}}\right)(30 \text{ m/s}) = -0.20 \text{ m/s}$$

Bob's recoil *speed* is 0.20 m/s.

**Assess:** Since the rock has forward momentum, Bob's momentum is backward. This makes the total momentum zero.

**9.23. Model:** We assume that the momentum is conserved in the collision.

**Visualize:** Please refer to Figure EX9.23.

**Solve:** The conservation of momentum equation yields

$$(p_{fx})_1 + (p_{fx})_2 = (p_{ix})_1 + (p_{ix})_2 \Rightarrow (p_{fx})_1 + 0 \text{ kg m/s} = 2 \text{ kg m/s} - 4 \text{ kg m/s} \Rightarrow (p_{fx})_1 = -2 \text{ kg m/s}$$

$$(p_{fy})_1 + (p_{fy})_2 = (p_{iy})_1 + (p_{iy})_2 \Rightarrow (p_{fy})_1 - 1 \text{ kg m/s} = 2 \text{ kg m/s} + 1 \text{ kg m/s} \Rightarrow (p_{fy})_1 = 4 \text{ kg m/s}$$

Thus, the final momentum of particle 1 is $(-2\hat{i} + 4\hat{j})$ kg m/s.

**9.29. Model:** Model the ball as a particle that is subjected to an impulse when it is in contact with the floor. We will also use constant-acceleration kinematic equations. Ignore any forces other than the interaction between the floor and the ball during the collision in the impulse approximation.

**Visualize:**

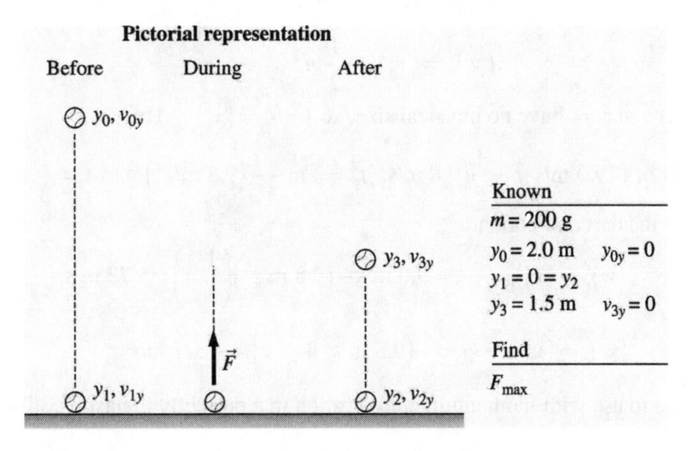


**Solve:** To find the ball's velocity just before and after it hits the floor:

$$v_{1y}^2 = v_{0y}^2 + 2a_y(y_1 - y_0) = 0 \text{ m}^2/\text{s}^2 + 2(-9.8 \text{ m/s}^2)(0 - 2.0 \text{ m}) \Rightarrow v_{1y} = -6.261 \text{ m/s}$$

$$v_{3y}^2 = v_{2y}^2 + 2a_y(y_3 - y_2) \Rightarrow 0 \text{ m}^2/\text{s}^2 = v_{2y}^2 + 2(-9.8 \text{ m/s}^2)(1.5 \text{ m} - 0 \text{ m}) \Rightarrow v_{2y} = 5.422 \text{ m/s}$$

The force exerted by the floor on the ball can be found from the impulse-momentum theorem:

$$mv_{2y} = mv_{1y} + \int F dt = mv_{1y} + \text{area under the force curve}$$

$$\Rightarrow (0.200 \text{ kg})(5.422 \text{ m/s}) = -(0.200 \text{ kg})(6.261 \text{ m/s}) + \tfrac{1}{2} F_{max}(5.0 \times 10^{-3} \text{ s})$$

$$\Rightarrow F_{max} = 9.3 \times 10^2 \text{ N}$$

**Assess:** A maximum force of $9.3 \times 10^2$ N exerted by the floor is reasonable.

**9.35. Model:** The dart and cork are particles in free fall until they have a perfectly inelastic head-on collision. Ignore air friction.

**Visualize:**

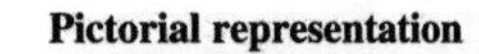


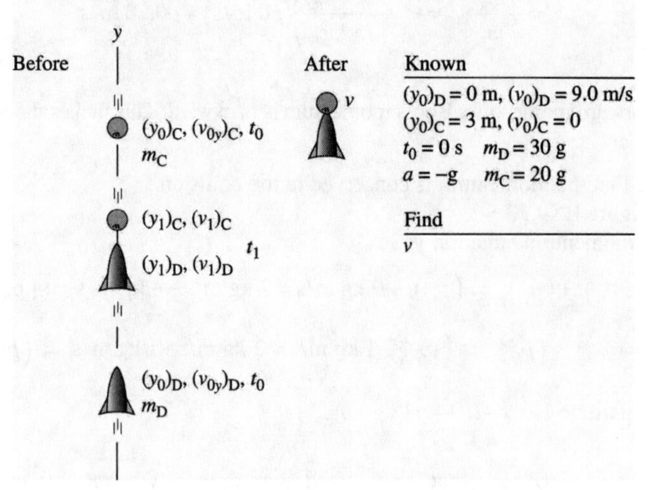


**Solve:** The positions of the dart and cork just before the collision are

$$(y_1)_D = (y_0)_D + (v_0)_D t_1 - \frac{1}{2} g t_1^2$$

$$(y_1)_C = (y_0)_C - \frac{1}{2} g t_1^2$$

In the particle model, the dart and cork have no physical size, so $(y_1)_D = (y_1)_C$. Hence

$$0 \text{ m} + (9.0 \text{ m/s}) t_1 - \frac{1}{2}(9.8 \text{ m/s}^2) t_1^2 = 3 \text{ m} - \frac{1}{2}(9.8 \text{ m/s}^2) t_1^2 \Rightarrow t_1 = \frac{1}{3} \text{ s}$$

At this time, the velocities of the dart and cork are

$$(v_1)_D = (v_0)_D - g t_1 = 9.0 \text{ m/s} - (9.8 \text{ m/s}^2)\left(\frac{1}{3} \text{ s}\right) = 5.73 \text{ m/s}$$

$$(v_1)_C = (v_0)_C - g t_1 = -(9.8 \text{ m/s}^2)\left(\frac{1}{3} \text{ s}\right) = -3.27 \text{ m/s}$$

These are the initial velocities to use with momentum conservation in a perfectly inelastic collision.

$$p_{fy} = p_{iy}$$

$$(m_D + m_C) v = m_D (v_1)_D + m_C (v_1)_C$$

Thus

$$v = \frac{(0.030 \text{ kg})(5.73 \text{ m/s}) + (0.020 \text{ kg})(-3.27 \text{ m/s})}{(0.030 \text{ kg} + 0.20 \text{ kg})} = 2.13 \text{ m/s, up}$$

**Assess:** The heavier, faster upward-going dart has more momentum than the falling cork, so the total momentum is upwards.

**9.39. Model:** This problem deals with a case that is the opposite of a collision. Our system is comprised of three coconut pieces that are modeled as particles. During the blow up or "explosion," the total momentum of the system is conserved in the *x*-direction and the *y*-direction.
**Visualize:**

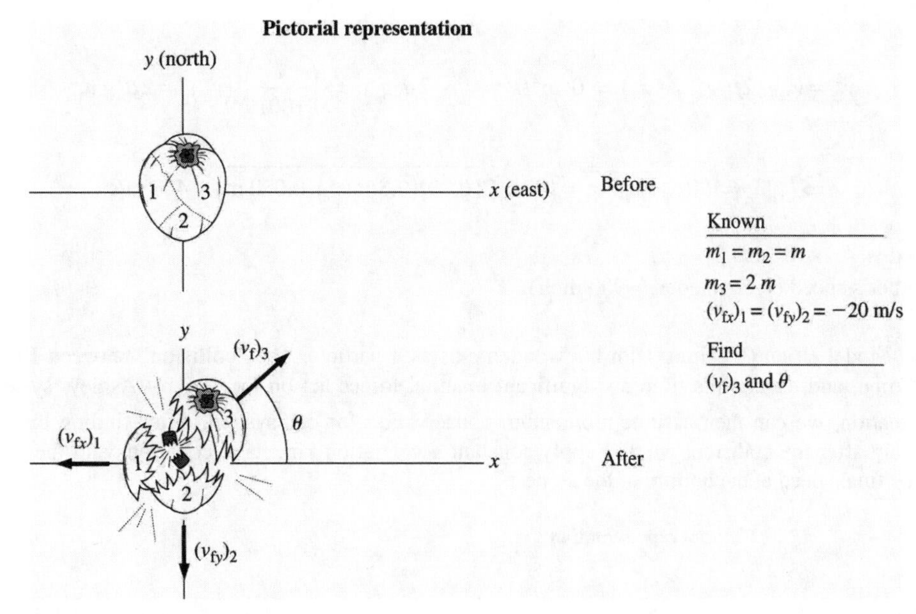


**Solve:** The initial momentum is zero. From $p_{fx} = p_{ix}$, we get

$$+m_1(v_{fx})_1 + m_3(v_f)_3\cos\theta = 0 \text{ kg m/s} \Rightarrow (v_f)_3\cos\theta = \frac{-m_1(v_{fx})_1}{m_3} = \frac{-m(-20 \text{ m/s})}{2m} = 10 \text{ m/s}$$

From $p_{fx} = p_{ix}$, we get

$$+m_2(v_{fy})_2 + m_3(v_f)_3\sin\theta = 0 \text{ kg m/s} \Rightarrow (v_f)_3\sin\theta = \frac{-m_2(v_{fy})_2}{m_3} = \frac{-m(-20 \text{ m/s})}{2m} = 10 \text{ m/s}$$

$$\Rightarrow (v_f)_3 = \sqrt{(10 \text{ m/s})^2 + (10 \text{ m/s})^2} = 14.1 \text{ m/s} \quad \theta = \tan^{-1}(1) = 45°$$

The velocity is 14.1 m/s at 45° east of north.

**9.41. Model:** This is a two-part problem. First, we have an inelastic collision between the wood block and the bullet. The bullet and the wood block are an isolated system. Since any external force acting during the collision is not going to be significant (the impulse approximation), the momentum of the system will be conserved. The second part involves the dynamics of the block + bullet sliding on the wood table. We treat the block and the bullet as particles.
**Visualize:**

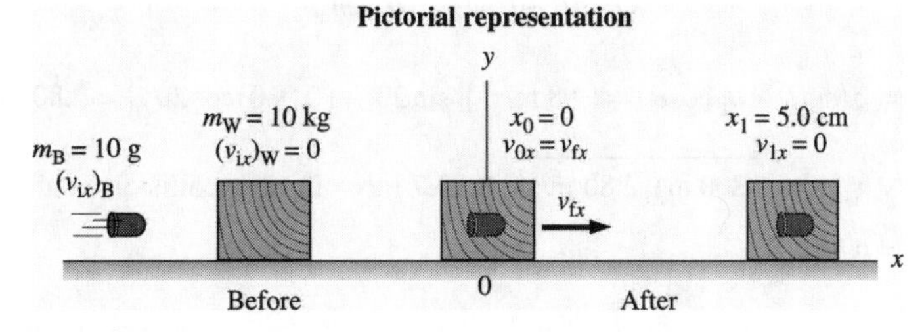


**Solve:** The equation $p_{fx} = p_{ix}$ gives

$$(m_B + m_W)v_{fx} = m_B(v_{ix})_B + m_W(v_{ix})_W$$

$$\Rightarrow (0.010 \text{ kg} + 10 \text{ kg})v_{fx} = (0.010 \text{ kg})(v_{ix})_B + (10 \text{ kg})(0 \text{ m/s}) \Rightarrow v_{fx} = \frac{1}{1001}(v_{ix})_B$$

From the model of kinetic friction,

$$f_k = -\mu_k n = -\mu_k (m_B + m_W) g = (m_B + m_W) a_x \Rightarrow a_x = -\mu_k g$$

Using the kinematic equation $v_{1x}^2 = v_{2x}^2 + 2a_x(x_1 - x_0)$,

$$v_{1x}^2 = v_{2x}^2 - 2\mu_k g(x_1 - x_0) \Rightarrow 0\ \text{m}^2/\text{s}^2 = v_{fx}^2 - 2\mu_k g\ x_1 \Rightarrow \left(\frac{1}{1001}\right)^2 (v_{ix})_B^2 = 2\mu_k g\ x_1$$

$$\Rightarrow (v_{ix})_B = 1001\sqrt{2\mu_k g\ x_1} = 1001\sqrt{2(0.20)(9.8\ \text{m/s}^2)(0.050\ \text{m})} = 443\ \text{m/s}$$

The bullet's speed is $4.4 \times 10^2$ m/s.

**Assess:** The bullet's speed is reasonable ($\approx$900 mph).

**9.49. Model:** Model Brian (B) along with his wooden skis as a particle. The "collision" between Brian and Ashley lasts for a short time, and during this time no significant external forces act on the Brian + Ashley system. Within the impulse approximation, we can then assume momentum conservation for our system. After finding the velocity of the system immediately after the collision, we will apply constant-acceleration kinematic equations and the model of kinetic friction to find the final speed at the bottom of the slope.

**Visualize:**

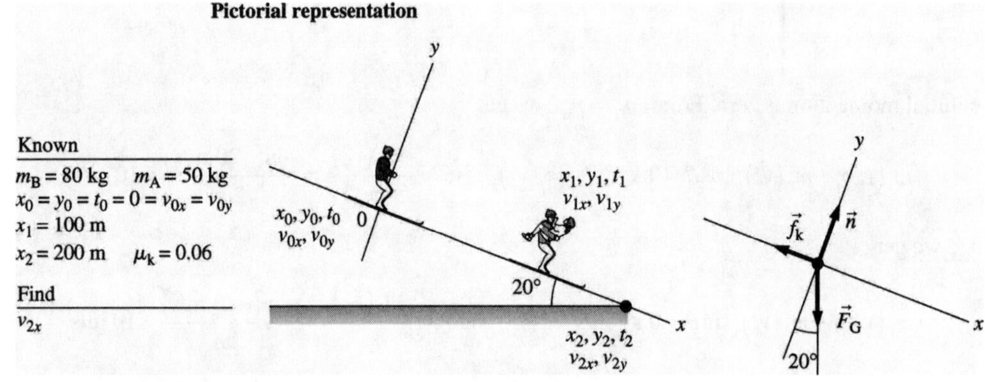


**Solve:** Brian skiing down for 100 m:

$$(v_{1x})_B^2 = (v_{0x})_B^2 + 2a_x(x_{1B} - x_{0B}) = 0\ \text{m}^2/\text{s}^2 + 2a_x(100\ \text{m} - 0\ \text{m}) \Rightarrow (v_{1x})_B = \sqrt{(200\ \text{m})a_x}$$

To obtain $a_x$, we apply Newton's second law to Brian in the *x* and *y* directions as follows:

$$\sum(F_{\text{on B}})_x = w_B \sin\theta - f_k = m_B a_x \quad \sum(F_{\text{on B}})_y = n - w_B\cos\theta = 0\ \text{N} \Rightarrow n = w\cos\theta$$

From the model of kinetic friction, $f_k = \mu_k n = \mu_k w_B \cos\theta$. The *x*-equation thus becomes

$$w_B \sin\theta - \mu_k w_B \cos\theta = m_B a_x$$

$$\Rightarrow a_x = g(\sin\theta - \mu_k\cos\theta) = (9.8\ \text{m/s}^2)[\sin 20° - (0.060)\cos 20°] = 2.80\ \text{m/s}^2$$

Using this value of $a_x$, $(v_{1x})_B = \sqrt{(200\ \text{m})(2.80\ \text{m/s}^2)} = 23.7$ m/s. In the collision with Ashley the conservation of momentum equation $p_{fx} = p_{ix}$ is

$$(m_B + m_A)v_{2x} = m_B(v_{1x})_B \Rightarrow v_{2x} = \frac{m_B}{m_B + m_A}(v_{1x})_B = \frac{80\ \text{kg}}{80\ \text{kg} + 50\ \text{kg}}(23.66\ \text{m/s}) = 14.56\ \text{m/s}$$

Brian + Ashley skiing down the slope:

$$v_{3x}^2 = v_{2x}^2 + 2a_x(x_3 - x_2) = (14.56\ \text{m/s})^2 + 2(2.80\ \text{m/s}^2)(100\ \text{m}) \Rightarrow v_{3x} = 27.8\ \text{m/s}$$

That is, Brian + Ashley arrive at the bottom of the slope with a speed of 27.8 m/s. Note that we have used the same value of $a_x$ in the first and the last parts of this problem. This is because $a_x$ is independent of mass.

**Assess:** A speed of approximately 60 mph on a ski slope of 200 m length and 20° slope is reasonable.

**9.51. Model:** This is an isolated system, so momentum is conserved in the explosion. Momentum is a *vector* quantity, so the direction of the initial velocity vector $\vec{v}_1$ establishes the direction of the momentum vector. The final momentum vector, after the explosion, must still point in the +*x*-direction. The two known pieces continue to move along this line and have no *y*-components of momentum. The missing third piece cannot have a *y*-component of momentum if momentum is to be conserved, so it must move along the *x*-axis—either straight forward or straight backward. We can use conservation laws to find out.

**Visualize:**

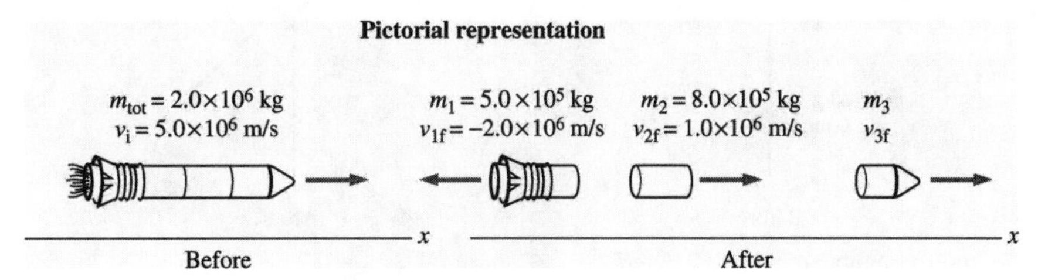


**Solve:** From the conservation of mass, the mass of piece 3 is

$$m_3 = m_{\text{total}} - m_1 - m_2 = 7.0\times10^5 \text{ kg}$$

To conserve momentum along the *x*-axis, we require

$$\left[p_i = m_{\text{total}}v_i\right] = \left[p_f = p_{1f} + p_{2f} + p_{3f} = m_1v_{1f} + m_2v_{2f} + p_{3f}\right]$$

$$\Rightarrow p_{3f} = m_{\text{total}}v_i - m_1v_{1f} - m_2v_{2f} = +1.02\times10^{13} \text{ kg m/s}$$

Because $p_{3f} > 0$, the third piece moves in the +*x*-direction, that is, straight forward. Because we know the mass $m_3$, we can find the velocity of the third piece as follows:

$$v_{3f} = \frac{p_{3f}}{m_3} = \frac{1.02\times10^{13} \text{ kg m/s}}{7.0\times10^5 \text{ kg}} = 1.46\times10^7 \text{ m/s}$$

**9.57. Model:** The neutron's decay is an "explosion" of the neutron into several pieces. The neutron is an isolated system, so its momentum should be conserved. The observed decay products, the electron and proton, move in opposite directions.

**Visualize:**

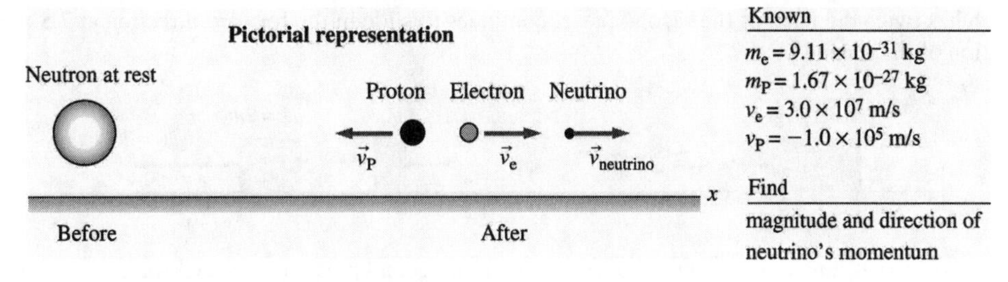


**Solve:** **(a)** The initial momentum is $p_{ix} = 0$ kg m/s. The final momentum $p_{fx} = m_ev_e + m_pv_p$ is

$$2.73\times10^{-23} \text{ kg m/s} - 1.67\times10^{-22} \text{ kg m/s} = -1.40\times10^{-22} \text{ kg m/s}$$

No, momentum does not seem to be conserved.

**(b)** and **(c)** If the neutrino is needed to conserve momentum, then $p_e + p_P + p_{\text{neutrino}} = 0$ kg m/s. This requires

$$p_{\text{neutrino}} = -(p_e + p_P) = +1.40\times10^{-22} \text{ kg m/s}$$

The neutrino must "carry away" $1.40\times10^{-22}$ kg m/s of momentum in the same direction as the electron.

**9.59. Model:** Model the three balls of clay as particle 1 (moving north), particle 2 (moving west), and particle 3 (moving southeast). The three stick together during their collision, which is perfectly inelastic. The momentum of the system is conserved.

**Visualize:**

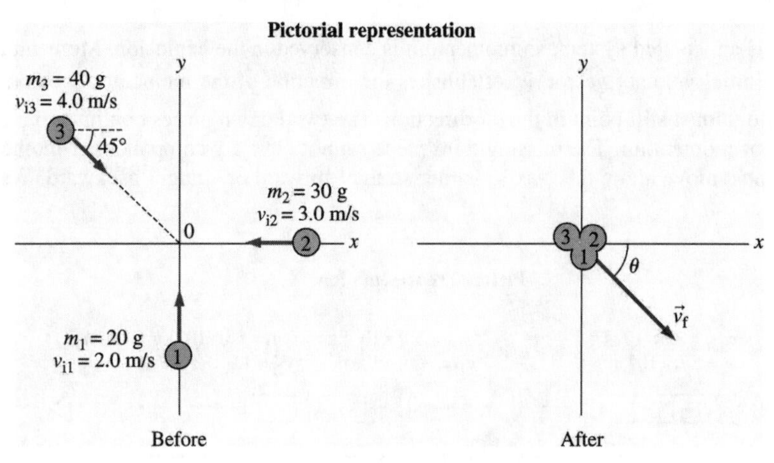


**Solve:** The three initial momenta are

$$\vec{p}_{i1} = m_1\vec{v}_{i1} = (0.020 \text{ kg})(2.0 \text{ m/s})\hat{j} = 0.040\hat{j} \text{ kg m/s}$$

$$\vec{p}_{i2} = m_2\vec{v}_{i2} = (0.030 \text{ kg})(-3.0 \text{ m/s } \hat{i}) = -0.090\hat{i} \text{ kg m/s}$$

$$\vec{p}_{i3} = m_3\vec{v}_{i3} = (0.040 \text{ kg})\left[(4.0 \text{ m/s})\cos 45°\hat{i} - (4.0 \text{ m/s})\sin 45°\hat{j}\right] = (0.113\hat{i} - 0.113\hat{j}) \text{ kg m/s}$$

Since $\vec{p}_f = \vec{p}_i = \vec{p}_{i1} + \vec{p}_{i2} + \vec{p}_{i3}$, we have

$$(m_1 + m_2 + m_3)\vec{v}_f = (0.023\hat{i} - 0.073\hat{j}) \text{ kg m/s} \Rightarrow \vec{v}_f = (0.256\hat{i} - 0.811\hat{j}) \text{ m/s}$$

$$\Rightarrow v_f = \sqrt{(0.256 \text{ m/s})^2 + (-0.811 \text{ m/s})^2} = 0.85 \text{ m/s}$$

$$\theta = \tan^{-1}\frac{|v_{fy}|}{v_{fx}} = \tan^{-1}\frac{0.811}{0.256} = 72° \text{ below } +x$$

**9.65.** **(a)** A 150 g spring-loaded toy is sliding across a frictionless floor at 1.0 m/s. It suddenly explodes into two pieces. One piece, which has twice the mass of the second piece, continues to slide in the forward direction at 7.5 m/s. What is the speed and direction of the second piece?

**(b)**

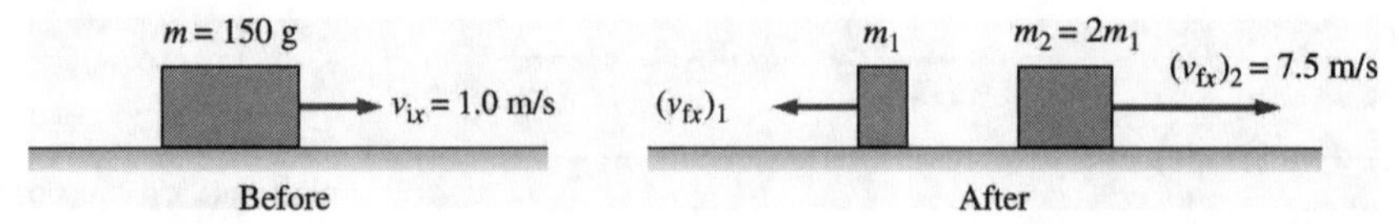


**(c)** The solution is $(v_{fx})_1 = -12$ m/s. The minus sign tells us that the second piece moves backward at 12 m/s.

# ENERGY

# 10

**10.3. Model:** Model the compact car (C) and the truck (T) as particles.
**Visualize:**

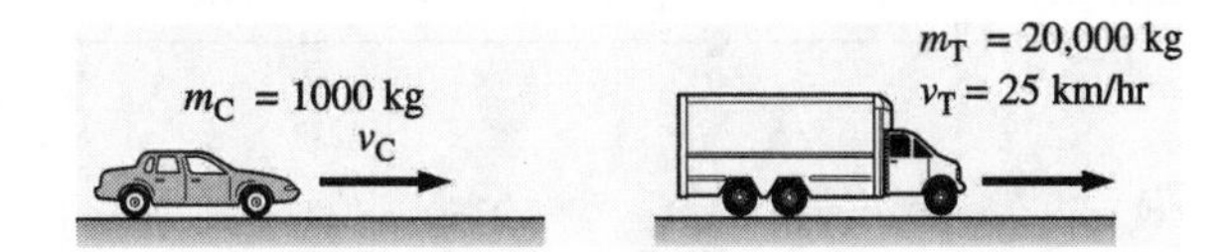


**Solve:** For the kinetic energy of the compact car and the kinetic energy of the truck to be equal,

$$K_C = K_T \Rightarrow \frac{1}{2}m_C v_C^2 = \frac{1}{2}m_T v_T^2 \Rightarrow v_C = \sqrt{\frac{m_T}{m_C}}v_T = \sqrt{\frac{20{,}000\ \text{kg}}{1000\ \text{kg}}}(25\ \text{km/hr}) = 112\ \text{km/hr}$$

**Assess:** A smaller mass needs a greater velocity for its kinetic energy to be the same as that of a larger mass.

**10.9. Model:** Model the skateboarder as a particle. Assuming that the track offers no rolling friction, the sum of the skateboarder's kinetic and gravitational potential energy does not change during his rolling motion.
**Visualize:**

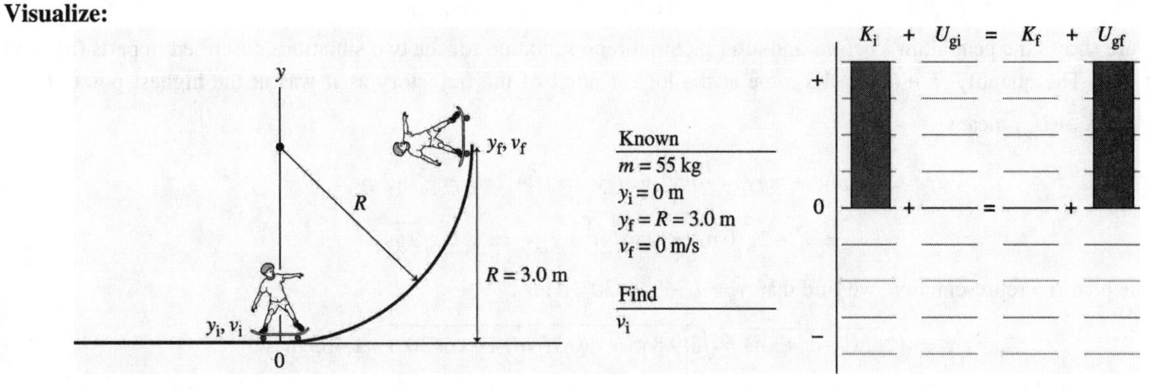


The vertical displacement of the skateboarder is equal to the radius of the track.
**Solve:** The quantity $K + U_g$ is the same at the upper edge of the quarter-pipe track as it was at the bottom. The energy conservation equation $K_f + U_{gf} = K_i + U_{gi}$ is

$$\frac{1}{2}mv_f^2 + mgy_f = \frac{1}{2}mv_i^2 + mgy_i \Rightarrow v_i^2 = v_f^2 + 2g(y_f - y_i)$$

$$v_i^2 = (0\ \text{m/s})^2 + 2(9.8\ \text{m/s}^2)(3.0\ \text{m} - 0\ \text{m}) = 58.8\ \text{m}^2/\text{s}^2 \Rightarrow v_i = 7.7\ \text{m/s}$$

**Assess:** Note that we did not need to know the skateboarder's mass, as is the case with free-fall motion.

**10.11. Model:** In the absence of frictional and air-drag effects, the sum of the kinetic and gravitational potential energy does not change as the pendulum swings from one side to the other.

**Visualize:**

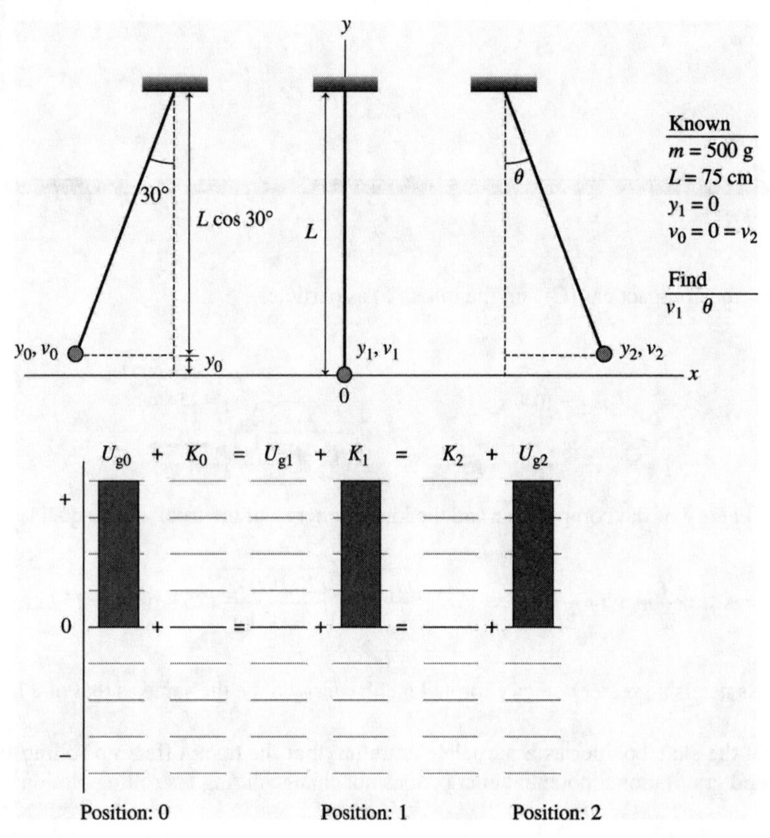


The figure shows the pendulum's before-and-after pictorial representation for the two situations described in parts (a) and (b).

**Solve:** **(a)** The quantity $K+U_g$ is the same at the lowest point of the trajectory as it was at the highest point. Thus, $K_1+U_{g1}=K_0+U_{g0}$ means

$$\frac{1}{2}mv_1^2+mgy_1=\frac{1}{2}mv_0^2+mgy_0 \Rightarrow v_1^2+2gy_1=v_0^2+2gy_0$$

$$\Rightarrow v_1^2+2g(0\text{ m})=(0\text{ m/s})^2+2gy_0 \Rightarrow v_1=\sqrt{2gy_0}$$

From the pictorial representation, we find that $y_0=L-L\cos 30°$. Thus,

$$v_1=\sqrt{2gL(1-\cos 30°)}=\sqrt{2(9.8\text{ m/s}^2)(0.75\text{ m})(1-\cos 30°)}=1.403\text{ m/s}$$

The speed at the lowest point is 1.40 m/s.

**(b)** Since the quantity $K+U_g$ does not change, $K_2+U_{g2}=K_1+U_{g1}$. We have

$$\frac{1}{2}mv_2^2+mgy_2=\frac{1}{2}mv_1^2+mgy_1 \Rightarrow y_2=\left(v_1^2-v_2^2\right)/2g$$

$$\Rightarrow y_2=[(1.403\text{ m/s})^2-(0\text{ m/s})^2]/(2\times 9.8\text{ m/s}^2)=0.100\text{ m}$$

Since $y_2=L-L\cos\theta$, we obtain

$$\cos\theta=\frac{L-y_2}{L}=\frac{(0.75\text{ m})-(0.10\text{ m})}{(0.75\text{ m})}=0.8667 \Rightarrow \theta=\cos^{-1}(0.8667)=30°$$

That is, the pendulum swings to the other side by 30°.

**Assess:** The swing angle is the same on either side of the rest position. This result is a consequence of the fact that the sum of the kinetic and gravitational potential energy does not change. This is shown as well in the energy bar chart in the figure.

**10.17. Model:** Assume that the spring is ideal and obeys Hooke's law. We also model the 5.0 kg mass as a particle.
**Visualize:** We will use the subscript s for the scale and sp for the spring.

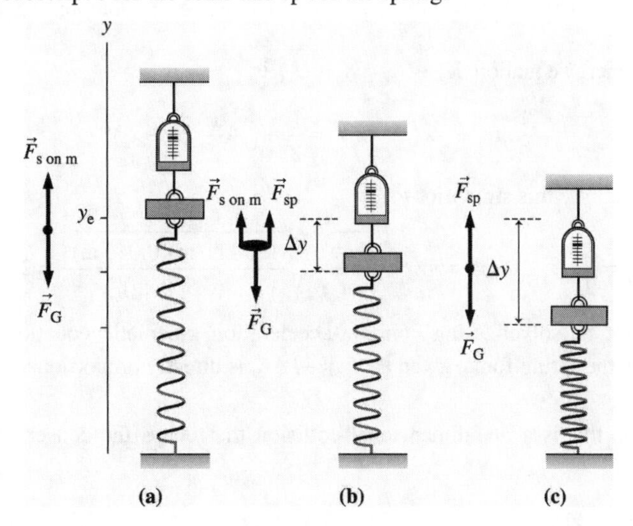


**Solve:** **(a)** The scale reads the upward force $F_{\text{s on m}}$ that it applies to the mass. Newton's second law gives

$$\sum(F_{\text{on m}})_y = F_{\text{s on m}} - F_G = 0 \Rightarrow F_{\text{s on m}} = F_G = mg = (5.0\text{ kg})(9.8\text{ m/s}^2) = 49\text{ N}$$

**(b)** In this case, the force is

$$\sum(F_{\text{on m}})_y = F_{\text{s on m}} + F_{\text{sp}} - F_G = 0 \Rightarrow 20\text{ N} + k\Delta y - mg = 0$$
$$\Rightarrow k = (mg - 20\text{ N})/\Delta y = (49\text{ N} - 20\text{ N})/0.02\text{ m} = 1450\text{ N/m}$$

The spring constant for the lower spring is $1.45\times10^3$ N/m.
**(c)** In this case, the force is

$$\sum(F_{\text{on m}})_y = F_{\text{sp}} - F_G = 0 \Rightarrow k\Delta y - mg = 0$$
$$\Rightarrow \Delta y = mg/k = (49\text{ N})/(1450\text{ N/m}) = 0.0338\text{ m} = 3.4\text{ cm}$$

**10.21. Model:** Assume an ideal spring that obeys Hooke's law. There is no friction, so the mechanical energy $K + U_s$ is conserved. Also model the book as a particle.
**Visualize:**

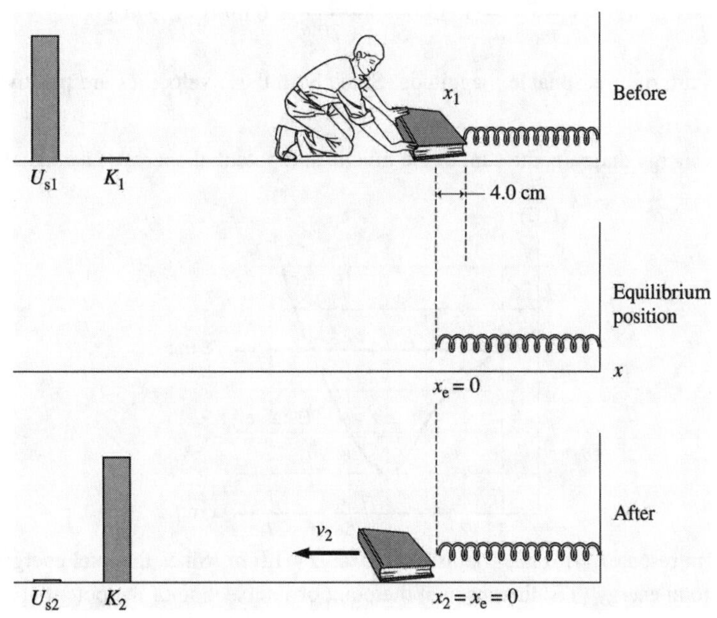

The figure shows a before-and-after pictorial representation. The compressed spring will push on the book until the spring has returned to its equilibrium length. We put the origin of our coordinate system at the equilibrium position of the free end of the spring. The energy bar chart shows that the potential energy of the compressed spring is entirely transformed into the kinetic energy of the book.

**Solve:** The conservation of energy equation $K_2 + U_{s2} = K_1 + U_{s1}$ is

$$\frac{1}{2}mv_2^2 + \frac{1}{2}k(x_2 - x_e)^2 = \frac{1}{2}mv_1^2 + \frac{1}{2}k(x_1 - x_e)^2$$

Using $x_2 = x_e = 0$ m and $v_1 = 0$ m/s, this simplifies to

$$\frac{1}{2}mv_2^2 = \frac{1}{2}k(x_1 - 0\text{ m})^2 \Rightarrow v_2 = \sqrt{\frac{kx_1^2}{m}} = \sqrt{\frac{(1250\text{ N/m})(0.040\text{ m})^2}{(0.500\text{ kg})}} = 2.0\text{ m/s}$$

**Assess:** This problem cannot be solved using constant-acceleration kinematic equations. The acceleration is not a constant in this problem, since the spring force, given as $F_s = -k\Delta x$, is directly proportional to $\Delta x$ or $|x - x_e|$.

**10.25. Model:** We assume this is a one-dimensional collision that obeys the conservation laws of momentum and mechanical energy.

**Visualize:**

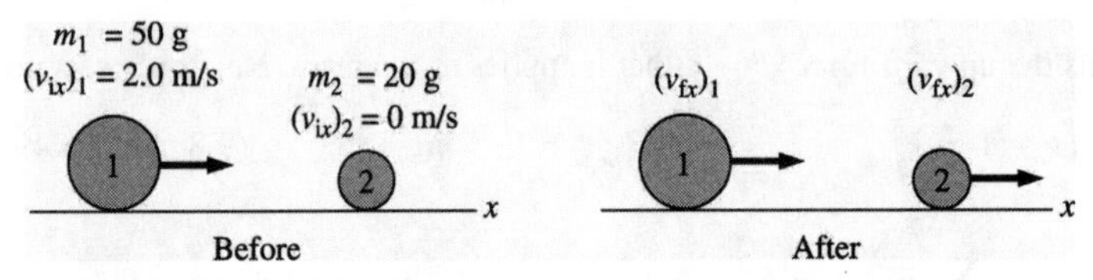


Note that momentum conservation alone is not sufficient to solve this problem because the two final velocities $(v_{fx})_1$ and $(v_{fx})_2$ are unknowns and can not be determined from one equation.

**Solve:** Momentum conservation: $m_1(v_{ix})_1 + m_2(v_{ix})_2 = m_1(v_{fx})_1 + m_2(v_{fx})_2$

Energy conservation: $\frac{1}{2}m_1(v_{ix})_1^2 + \frac{1}{2}m_2(v_{ix})_2^2 = \frac{1}{2}m_1(v_{fx})_1^2 + \frac{1}{2}m_2(v_{fx})_2^2$

These two equations can be solved for $(v_{fx})_1$ and $(v_{fx})_2$, as shown by Equations 10.39 through 10.43, to give

$$(v_{fx})_1 = \frac{m_1 - m_2}{m_1 + m_2}(v_{ix})_1 = \frac{50\text{ g} - 20\text{ g}}{50\text{ g} + 20\text{ g}}(2.0\text{ m/s}) = 0.86\text{ m/s}$$

$$(v_{fx})_2 = \frac{2m_1}{m_1 + m_2}(v_{ix})_1 = \frac{2(50\text{ g})}{50\text{ g} + 20\text{ g}}(2.0\text{ m/s}) = 2.9\text{ m/s}$$

**Assess:** These velocities are of a reasonable magnitude. Since both these velocities are positive, both balls move along the +*x*-direction.

**10.29. Model:** For an energy diagram, the sum of the kinetic and potential energy is a constant.

**Visualize:**

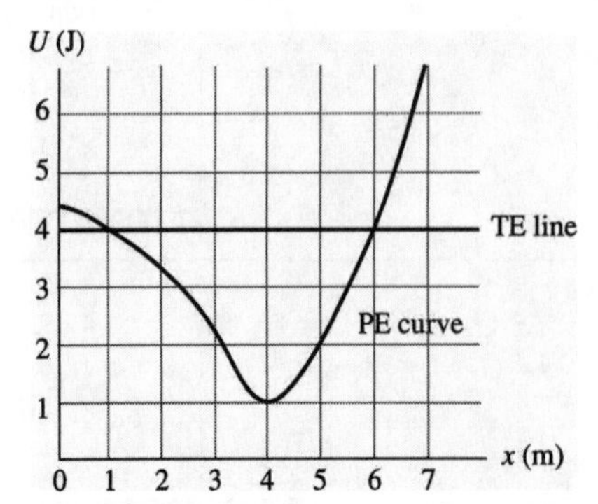


The particle is released from rest at $x = 1.0$ m. That is, $K = 0$ at $x = 1.0$ m. Since the total energy is given by $E = K + U$, we can draw a horizontal total energy (TE) line through the point of intersection of the potential energy curve (PE) and the

$x = 1.0$ m line. The distance from the PE curve to the TE line is the particle's kinetic energy. These values are transformed as the position changes, causing the particle to speed up or slow down, but the sum $K + U$ does not change.

**Solve:** **(a)** We have $E = 4.0$ J and this energy is a constant. For $x < 1.0$, $U > 4.0$ J and, therefore, $K$ must be negative to keep $E$ the same (note that $K = E - U$ or $K = 4.0 \text{ J} - U$). Since negative kinetic energy is unphysical, the particle can not move to the left. That is, the particle will move to the right of $x = 1.0$ m.

**(b)** The expression for the kinetic energy is $E - U$. This means the particle has maximum speed or maximum kinetic energy when $U$ is minimum. This happens at $x = 4.0$ m. Thus,

$$K_{\text{max}} = E - U_{\text{min}} = (4.0\text{ J}) - (1.0\text{ J}) = 3.0\text{ J} \qquad \frac{1}{2}mv_{\text{max}}^2 = 3.0\text{ J} \Rightarrow v_{\text{max}} = \sqrt{\frac{2(3.0\text{ J})}{m}} = \sqrt{\frac{8.0\text{ J}}{0.020\text{ kg}}} = 17.3\text{ m/s}$$

The particle possesses this speed at $x = 4.0$ m.

**(c)** The total energy (TE) line intersects the potential energy (PE) curve at $x = 1.0$ m and $x = 6.0$ m. These are the turning points of the motion.

**10.31. Model:** For an energy diagram, the sum of the kinetic and potential energy is a constant.

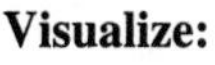

**Visualize:**

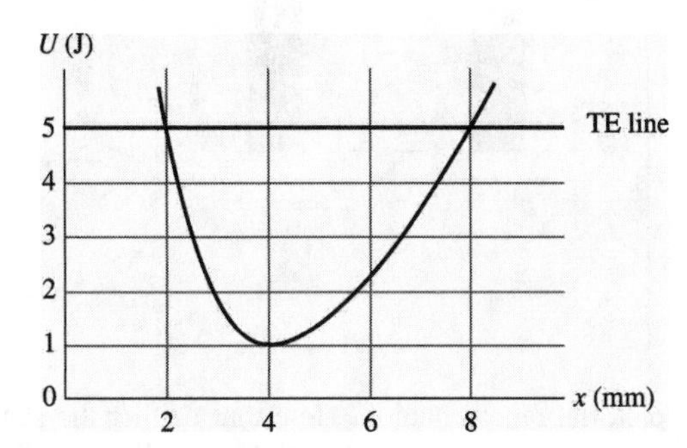


Since the particle oscillates between $x = 2.0$ mm and $x = 8.0$ mm, the speed of the particle is zero at these points. That is, for these values of $x$, $E = U = 5.0$ J, which defines the total energy (TE) line. The distance from the potential energy (PE) curve to the TE line is the particle's kinetic energy. These values are transformed as the position changes, but the sum $K + U$ does not change.

**Solve:** The equation for total energy $E = U + K$ means $K = E - U$, so that $K$ is maximum when $U$ is minimum. We have

$$K_{\text{max}} = \frac{1}{2}mv_{\text{max}}^2 = 5.0\text{ J} - U_{\text{min}}$$

$$\Rightarrow v_{\text{max}} = \sqrt{2(5.0\text{ J} - U_{\text{min}})/m} = \sqrt{2(5.0\text{ J} - 1.0\text{ J})/0.0020\text{ kg}} = 63\text{ m/s}$$

**10.33. Model:** Model your vehicle as a particle. Assume zero rolling friction, so that the sum of your kinetic and gravitational potential energy does not change as the vehicle coasts down the hill.

**Visualize:**

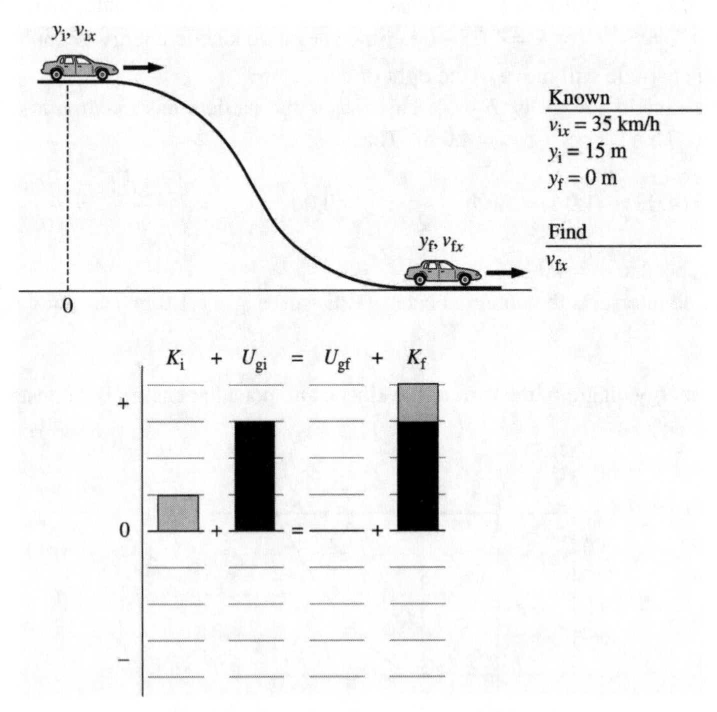


The figure shows a before-and-after pictorial representation. Note that neither the shape of the hill nor the angle of the downward slope is given, since these are not needed to solve the problem. All we need is the change in potential energy as you and your vehicle descend to the bottom of the hill. Also note that

$$35 \text{ km/hr} = (35{,}000 \text{ m}/3600 \text{ s}) = 9.722 \text{ m/s}$$

**Solve:** Using $y_f = 0$ and the equation $K_i + U_{gi} = K_f + U_{gf}$ we get

$$\frac{1}{2}mv_i^2 + mgy_i = \frac{1}{2}mv_f^2 + mgy_f \Rightarrow v_i^2 + 2gy_i = v_f^2$$

$$\Rightarrow v_f = \sqrt{v_i^2 + 2gy_i} = \sqrt{(9.722 \text{ m/s})^2 + 2(9.8 \text{ m/s}^2)(15 \text{ m})} = 19.7 \text{ m/s} = 71 \text{ km/h}$$

You are driving over the speed limit. Yes, you will get a ticket.

**Assess:** A speed of 19.7 m/s or 71 km/h at the bottom of the hill, when your speed at the top of the hill was 35 km/s, is reasonable. From the energy bar chart, we see that the initial potential energy is completely transformed into the final kinetic energy.

**10.37. Model:** Assume that the rubber band behaves similar to a spring. Also, model the rock as a particle.

**Visualize:**

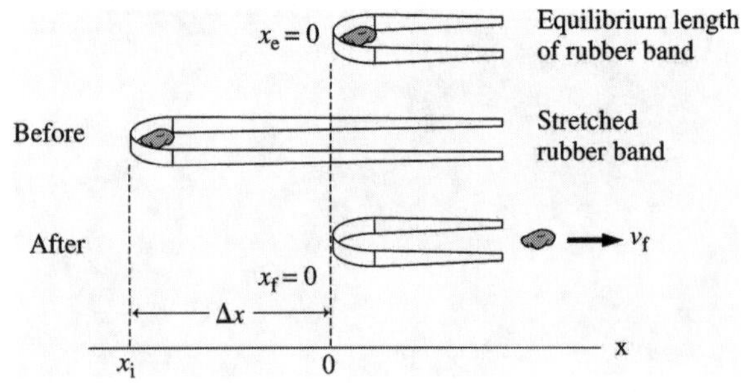


Please refer to Figure P10.37.

**Solve:** **(a)** The rubber band is stretched to the left since a positive spring force on the rock due to the rubber band results from a negative displacement of the rock. That is, $(F_{sp})_x = -kx$, where $x$ is the rock's displacement from the equilibrium position due to the spring force $F_{sp}$.

**(b)** Since the $F_{sp}$ versus $x$ graph is linear with a negative slope and can be expressed as $F_{sp} = -kx$, the rubber band obeys Hooke's law.

**(c)** From the graph, $|\Delta F_{sp}| = 20$ N for $|\Delta x| = 10$ cm. Thus,

$$k = \frac{|\Delta F_{sp}|}{|\Delta x|} = \frac{20\text{ N}}{0.10\text{ m}} = 200\text{ N/m} = 2.0\times10^2\text{ N/m}$$

**(d)** The conservation of energy equation $K_f + U_{sf} = K_i + U_{si}$ for the rock is

$$\frac{1}{2}mv_f^2 + \frac{1}{2}kx_f^2 = \frac{1}{2}mv_i^2 + \frac{1}{2}kx_i^2 \Rightarrow \frac{1}{2}mv_f^2 + \frac{1}{2}k(0\text{ m})^2 = \frac{1}{2}m(0\text{ m/s})^2 + \frac{1}{2}kx_i^2$$

$$v_f = \sqrt{\frac{k}{m}}x_i = \sqrt{\frac{200\text{ N/m}}{0.050\text{ kg}}}(0.30\text{ m}) = 19.0\text{ m/s}$$

**Assess:** Note that $x_i$ is $\Delta x$, which is the displacement relative to the equilibrium position, and $x_f$ is the equilibrium position of the rubber band, which is equal to zero.

**10.43. Model:** Model the marble and the steel ball as particles. We will assume an elastic collision between the marble and the ball, and apply the conservation of momentum and the conservation of energy equations. We will also assume zero rolling friction between the marble and the incline.

**Visualize:**

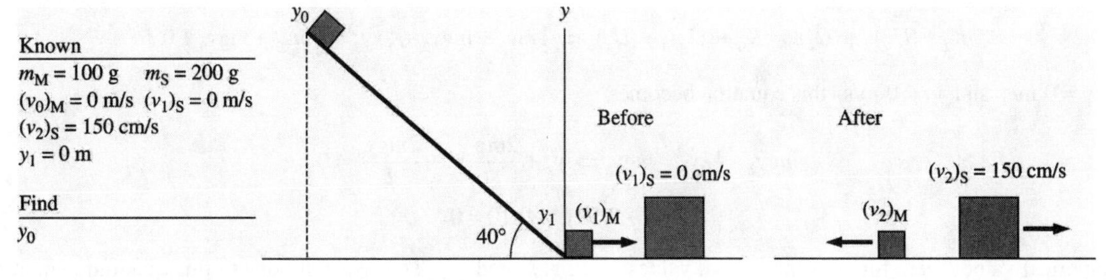


This is a two-part problem. In the first part, we will apply the conservation of energy equation to find the marble's speed as it exits onto a horizontal surface. We have put the origin of our coordinate system on the horizontal surface just where the marble exits the incline. In the second part, we will consider the elastic collision between the marble and the steel ball.

**Solve:** The conservation of energy equation $K_1 + U_{g1} = K_0 + U_{g0}$ gives us:

$$\frac{1}{2}m_M(v_1)_M^2 + m_M g y_1 = \frac{1}{2}m_M(v_0)_M^2 + m_M g y_0$$

Using $(v_0)_M = 0$ m/s and $y_1 = 0$ m, we get $\frac{1}{2}(v_1)_M^2 = gy_0 \Rightarrow (v_1)_M = \sqrt{2gy_0}$. When the marble collides with the steel ball, the elastic collision gives the ball velocity

$$(v_2)_S = \frac{2m_M}{m_M + m_S}(v_1)_M = \frac{2m_M}{m_M + m_S}\sqrt{2gy_0}$$

Solving for $y_0$ gives

$$y_0 = \frac{1}{2g}\left[\frac{m_M + m_S}{2m_M}(v_2)_S\right]^2 = 0.258\text{ m} = 25.8\text{ cm}$$

**10.47. Model:** Assume an ideal spring that obeys Hooke's law. There is no friction, so the mechanical energy $K+U_g+U_s$ is conserved.

**Visualize:**

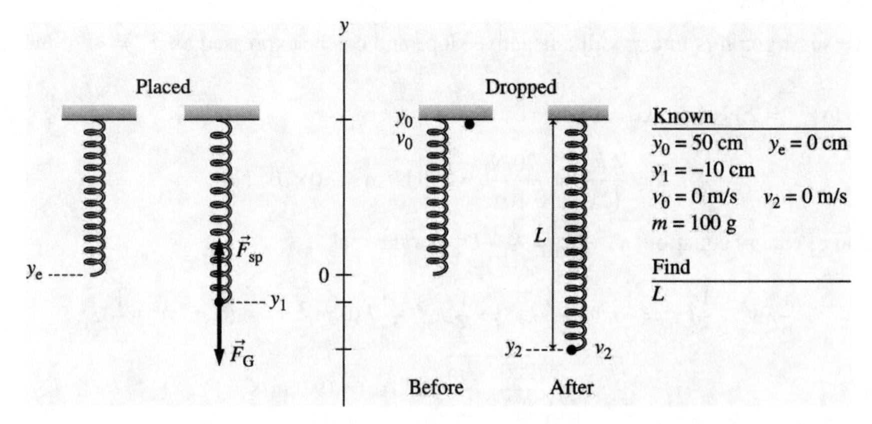


Place the origin of the coordinate system at the end of the unstretched spring, making $y_e = 0$ m.

**Solve:** The clay is in static equilibrium while resting in the pan. The net force on it is zero. We can start by using this to find the spring constant.

$$F_{sp} = F_G \Rightarrow -k(y_1 - y_e) = -ky_1 = mg \Rightarrow k = -\frac{mg}{y_1} = -\frac{(0.10 \text{ kg})(9.8 \text{ m/s}^2)}{-0.10 \text{ m}} = 9.8 \text{ N/m}$$

Now apply conservation of energy. Initially, the spring is unstretched and the clay ball is at the ceiling. At the end, the spring has maximum stretch and the clay is instantaneously at rest. Thus

$$K_2 + (U_g)_2 + (U_s)_2 = K_0 + (U_g)_0 + (U_s)_0 \Rightarrow \tfrac{1}{2}mv_2^2 + mgy_2 + \tfrac{1}{2}ky_2^2 = \tfrac{1}{2}mv_0^2 + mgy_0 + 0 \text{ J}$$

Since $v_0 = 0$ m/s and $v_2 = 0$ m/s, this equation becomes

$$mgy_2 + \tfrac{1}{2}ky_2^2 = mgy_0 \Rightarrow y_2^2 + \frac{2mg}{k}y_2 - \frac{2mgy_0}{k} = 0$$

$$y_2^2 + 0.20y_2 - 0.10 = 0$$

The numerical values were found using known values of $m$, $g$, $k$, and $y_0$. The two solutions to this quadratic equation are $y_2 = 0.231$ m and $y_2 = -0.432$ m. The point we're looking for is below the origin, so we need the negative root. The distance of the pan from the ceiling is

$$L = |y_2| + 50 \text{ cm} = 93 \text{ cm}$$

**10.53. Model:** Because the track is frictionless, the sum of the kinetic and gravitational potential energy does not change during the car's motion.

**Visualize:**

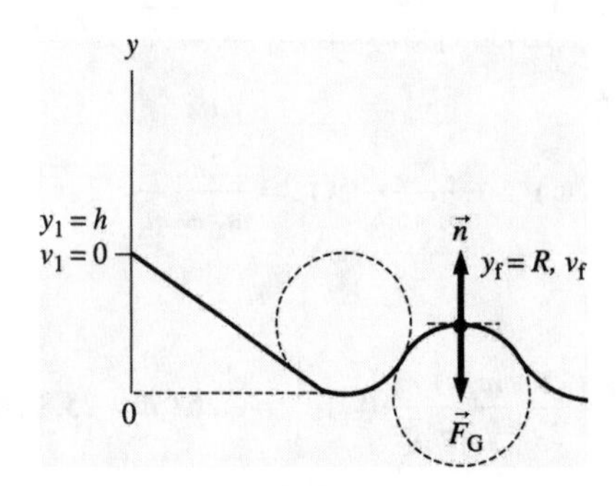


We place the origin of the coordinate system at the ground level directly below the car's starting position. This is a two-part problem. If we first find the maximum speed at the top of the hill, we can use energy conservation to find the maximum initial height.

**Solve:** Because its motion is circular, at the top of the hill the car has a downward-pointing centripetal acceleration $\vec{a}_c = -(mv^2/r)\hat{j}$. Newton's second law at the top of the hill is

$$(F_{net})_y = n_y + (F_G)_y = n - mg = m(a_c)_y = -\frac{mv^2}{R} \Rightarrow n = mg - \frac{mv^2}{R} = m\left(g - \frac{v^2}{R}\right)$$

If $v = 0$ m/s, $n = mg$ as expected in static equilibrium. As $v$ increases, $n$ gets smaller—the weight of the car and passengers decreases as they go over the top. But $n$ has to remain positive for the car to be on the track, so the maximum speed $v_{max}$ occurs when $n \to 0$. We see that $v_{max} = \sqrt{gR}$. Now we can use energy conservation to relate the top of the hill to the starting height:

$$K_f + U_f = K_i + U_i \Rightarrow \frac{1}{2}mv_f^2 + mgy_f = \frac{1}{2}mv_i^2 + mgy_i \Rightarrow \frac{1}{2}mgR + mgR = 0\text{ J} + mgh_{max}$$

where we used $v_f = v_{max}$ and $y_f = R$. Solving for $h_{max}$ gives $h_{max} = \frac{3}{2}R$.

**(b)** If $R = 10$ m, then $h_{max} = 15$ m.

**10.55. Model:** Model Lisa (L) and the bobsled (B) as particles. We will assume the ramp to be frictionless, so that the mechanical energy of the system (Lisa + bobsled + spring) is conserved. Furthermore, during the collision, as Lisa leaps onto the bobsled, the momentum of the Lisa + bobsled system is conserved. We will also assume the spring to be an ideal one that obeys Hooke's law.

**Visualize:**

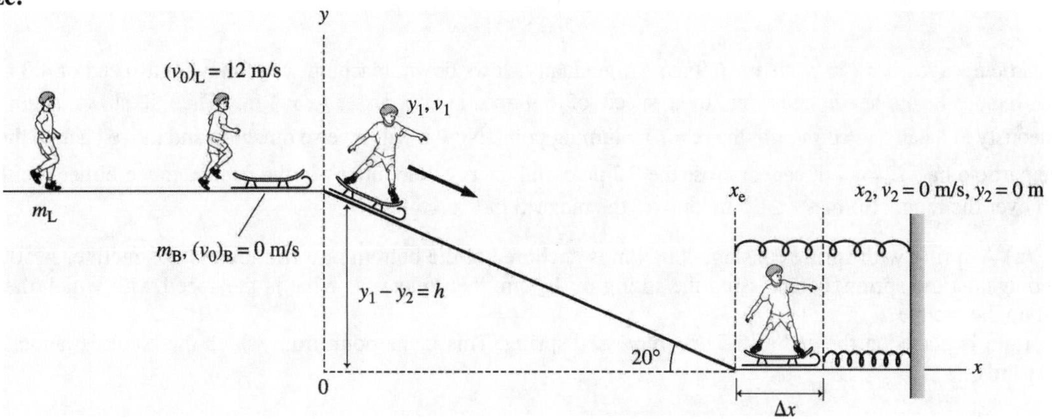


We place the origin of our coordinate system directly below the bobsled's initial position.

**Solve:** **(a)** Momentum conservation in Lisa's collision with bobsled states $p_1 = p_0$, or

$$(m_L + m_B)v_1 = m_L(v_0)_L + m_B(v_0)_B \Rightarrow (m_L + m_B)v_1 = m_L(v_0)_L + 0$$

$$\Rightarrow v_1 = \left(\frac{m_L}{m_L + m_B}\right)(v_0)_L = \left(\frac{40\text{ kg}}{40\text{ kg} + 20\text{ kg}}\right)(12\text{ m/s}) = 8.0\text{ m/s}$$

The energy conservation equation: $K_2 + U_{s2} + U_{g2} = K_1 + U_{s1} + U_{g1}$ is

$$\frac{1}{2}(m_L + m_B)v_2^2 + \frac{1}{2}k(x_2 - x_e)^2 + (m_L + m_B)gy_2 = \frac{1}{2}(m_L + m_B)v_1^2 + \frac{1}{2}k(x_e - x_e)^2 + (m_L + m_B)gy_1$$

Using $v_2 = 0$ m/s, $k = 2000$ N/m, $y_2 = 0$ m, $y_1 = (50\text{ m})\sin 20° = 17.1$ m, $v_1 = 8.0$ m/s, and $(m_L m_B) = 60$ kg, we get

$$0\text{ J} + \frac{1}{2}(2000\text{ N/m})(x_2 - x_e)^2 + 0\text{ J} = \frac{1}{2}(60\text{ kg})(8.0\text{ m/s})^2 + 0\text{ J} + (60\text{ kg})(9.8\text{ m/s}^2)(17.1\text{ m})$$

Solving this equation yields $(x_2 - x_e) = 3.5$ m.

**(b)** As long as the ice is slippery enough to be considered frictionless, we know from conservation of mechanical energy that the speed at the bottom depends only on the vertical descent $\Delta y$. Only the ramp's height $h$ is important, not its shape or angle.

**10.59. Model:** Use the model of the conservation of mechanical energy.

**Visualize:**

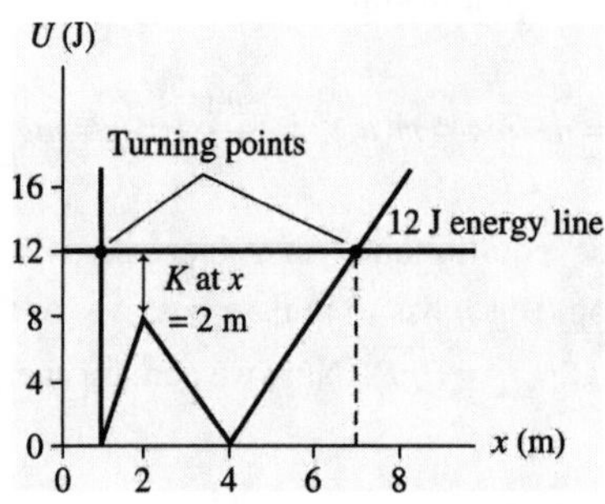


**Solve:** **(a)** The turning points occur where the total energy line crosses the potential energy curve. For $E = 12$ J, this occurs at the points $x = 1$ m and $x = 7$ m.

**(b)** The equation for kinetic energy $K = E - U$ gives the distance between the potential energy curve and total energy line. $U = 8$ J at $x = 2$ m, so $K = 12\text{ J} - 8\text{ J} = 4$ J. The speed corresponding to this kinetic energy is

$$v = \sqrt{\frac{2K}{m}} = \sqrt{\frac{2(4\text{ J})}{0.5\text{ kg}}} = 4.0\text{ m/s}$$

**(c)** Maximum speed occurs for minimum *U*. This occurs at $x = 1$ m and $x = 4$ m, where $U = 0$ J and $K = 12$ J. The speed at these two points is

$$v = \sqrt{\frac{2K}{m}} = \sqrt{\frac{2(12\text{ J})}{0.500\text{ kg}}} = 6.9\text{ m/s}$$

**(d)** The particle leaves $x = 1$ m with $v = 6.9$ m/s. It gradually slows down, reaching $x = 2$ m with a speed of 4.0 m/s. After $x = 2$ m, it speeds up again, returning to a speed of 6.9 m/s as it crosses $x = 4$ m. Then it slows again, coming instantaneously to a halt ($v = 0$ m/s) at the $x = 7$ m turning point. Now it will reverse direction and move back to the left.

**(e)** If the particle has $E = 4$ J it cannot cross the 8 J potential energy "mountain" in the center. It can either oscillate back and forth over the range $1.0\text{ m} \le x \le 1.5$ m or over the range $3\text{ m} \le x \le 5$ m.

**10.67.** **(a)** A spring with spring constant 400 N/m is anchored at the bottom of a frictionless 30° incline. A 500 g block is pressed against the spring, compressing the spring by 10 cm, then released. What is the speed with which the block is launched up the incline?

**(b)** The origin is placed at the end of the uncompressed spring. This is the point from which the block is launched as the spring expands.

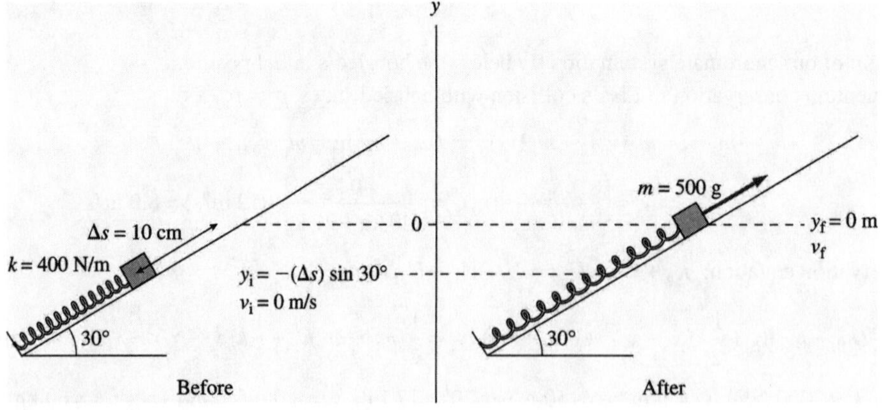


**(c)** Solving the energy conservation equation, we get $v_f = 2.6$ m/s.

# 11

# WORK

**11.7. Solve:** **(a)** $W = \vec{F} \cdot \Delta\vec{r} = (6.0\hat{\imath} - 3.0\hat{\jmath}) \cdot (2.0\hat{\imath})\ \text{N} \cdot \text{m} = (12.0\hat{\imath} \cdot \hat{\imath} - 3.0\hat{\jmath} \cdot \hat{\imath})\ \text{J} = 12.0\ \text{J}.$

**(b)** $W = \vec{F} \cdot \Delta\vec{r} = (6.0\hat{\imath} - 3.0\hat{\jmath}) \cdot (2.0\hat{\jmath})\ \text{N} \cdot \text{m} = (12.0\hat{\imath} \cdot \hat{\jmath} - 6.0\hat{\jmath} \cdot \hat{\jmath})\ \text{J} = -6.0\ \text{J}.$

**11.9. Model:** Use the work-kinetic energy theorem to find the net work done on the particle.
**Visualize:**

$\vec{F}$
$v_f = 30$ m/s
$v_i = 30$ m/s
$m = 20$ g
After
Before

**Solve:** From the work-kinetic energy theorem,

$$W = \Delta K = \frac{1}{2}mv_1^2 - \frac{1}{2}mv_0^2 = \frac{1}{2}m\left(v_1^2 - v_0^2\right) = \frac{1}{2}(0.020\ \text{kg})[(30\ \text{m/s})^2 - (-30\ \text{m/s})^2] = 0\ \text{J}$$

**Assess:** Negative work is done in slowing down the particle to rest, and an equal amount of positive work is done in bringing the particle to the original speed but in the opposite direction.

**11.11. Model:** Model the piano as a particle and use $W = \vec{F} \cdot \Delta\vec{r}$, where $W$ is the work done by the force $\vec{F}$ through the displacement $\Delta\vec{r}$.
**Visualize:**

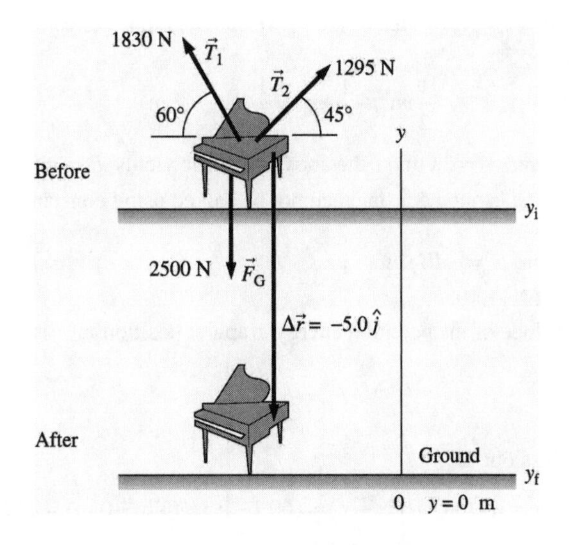


**Solve:** For the force $\vec{F}_G$:

$$W = \vec{F} \cdot \Delta\vec{r} = \vec{F}_G \cdot \Delta\vec{r} = (F_g)(\Delta r)\cos 0° = (2500\ \text{N})(5.00\ \text{m})(1) = 1.250 \times 10^4\ \text{J}$$

For the tension $\vec{T}_1$:

$$W = \vec{T}_1 \cdot \Delta\vec{r} = (T_1)(\Delta r)\cos(150°) = (1830 \text{ N})(5.00 \text{ m})(-0.8660) = -7.92\times10^3 \text{ J}$$

For the tension $\vec{T}_2$:

$$W = \vec{T}_2 \cdot \Delta\vec{r} = (T_2)(\Delta r)\cos(135°) = (1295 \text{ N})(5.00 \text{ m})(-0.7071) = -4.58\times10^3 \text{ J}$$

**Assess:** Note that the displacement $\Delta\vec{r}$ in all the above cases is directed downwards along $-\hat{j}$.

**11.13. Model:** Model the 2.0 kg object as a particle, and use the work-kinetic energy theorem.
**Visualize:** Please refer to Figure EX11.13. For each of the five intervals the velocity-versus-time graph gives the initial and final velocities. The mass of the object is 2.0 kg.
**Solve:** According to the work-kinetic energy theorem:

$$W = \Delta K = \frac{1}{2}mv_f^2 = \frac{1}{2}mv_i^2 = \frac{1}{2}m\left(v_f^2 - v_i^2\right)$$

$$\text{Interval AB: } v_i = 2 \text{ m/s}, \quad v_f = -2 \text{ m/s} \Rightarrow W = \frac{1}{2}(2.0 \text{ kg})[(-2 \text{ m/s})^2 - (2 \text{ m/s})^2] = 0 \text{ J}$$

$$\text{Interval BC: } v_i = -2 \text{ m/s}, \quad v_f = -2 \text{ m/s} \Rightarrow W = \frac{1}{2}(2.0 \text{ kg})[(-2 \text{ m/s})^2 - (-2 \text{ m/s})^2] = 0 \text{ J}$$

$$\text{Interval CD: } v_i = -2 \text{ m/s}, \quad v_f = 0 \text{ m/s} \Rightarrow W = \frac{1}{2}(2.0 \text{ kg})[(0 \text{ m/s})^2 - (-2 \text{ m/s})^2] = -4.0 \text{ J}$$

$$\text{Interval DE: } v_i = 0 \text{ m/s}, \quad v_f = 2 \text{ m/s} \Rightarrow W = \frac{1}{2}(2.0 \text{ kg})[(2 \text{ m/s})^2 - (0 \text{ m/s})^2] = +4.0 \text{ J}$$

$$\text{Interval EF: } v_i = 2 \text{ m/s}, \quad v_f = 1 \text{ m/s} \Rightarrow W = \frac{1}{2}(2.0 \text{ kg})[(1 \text{ m/s})^2 - (2 \text{ m/s})^2] = -3.0 \text{ J}$$

**Assess:** The work done is zero in intervals AB and BC. In the interval $\text{CD} + \text{DE}$ the total work done is zero. It is not whether $v$ is positive or negative that counts because $K \propto v^2$. What is important is the magnitude of $v$ and how $v$ changes.

**11.17. Model:** Use the work-kinetic energy theorem.
**Visualize:** Please refer to Figure EX11.17.
**Solve:** The work-kinetic energy theorem is

$$\Delta K = W = \int_{x_i}^{x_f} F_x\, dx = \text{area of the } F_x\text{-versus-}x \text{ graph between } x_i \text{ and } x_f$$

$$\frac{1}{2}mv_f^2 - \frac{1}{2}mv_i^2 = \frac{1}{2}(F_{max})(2 \text{ m})$$

Using $m = 0.500$ kg, $v_f = 6.0$ m/s, and $v_i = 2.0$ m/s, the above equation yields $F_{max} = 8.0$ N.
**Assess:** Problems in which the force is not a constant can not be solved using constant-acceleration kinematic equations.

**11.19. Model:** Use the definition $F_s = -dU/ds$.
**Visualize:** Please refer to Figure EX11.19.
**Solve:** $F_x$ is the negative of the slope of the potential energy graph at position $x$.

$$F_x = -\left(\frac{dU}{dx}\right)$$

Between $x = 0$ m and $x = 3$ m, the slope is

$$\text{slope} = (U_f - U_i)/(x_f - x_i) = (60 \text{ J} - 0 \text{ J})/(3 \text{ m} - 0 \text{ m}) = 20 \text{ N}$$

Thus, $F_x = -20$ N at $x = 1$ m. Between $x = 3$ m and $x = 5$ m, the slope is

$$\text{slope} = (U_f - U_i)/(x_f - x_i) = (0 \text{ J} - 60 \text{ J})/(5 \text{ m} - 3 \text{ m}) = -30 \text{ N}$$

Thus, $F_x = 30$ N at $x = 4$ m.

**11.25. Visualize:**

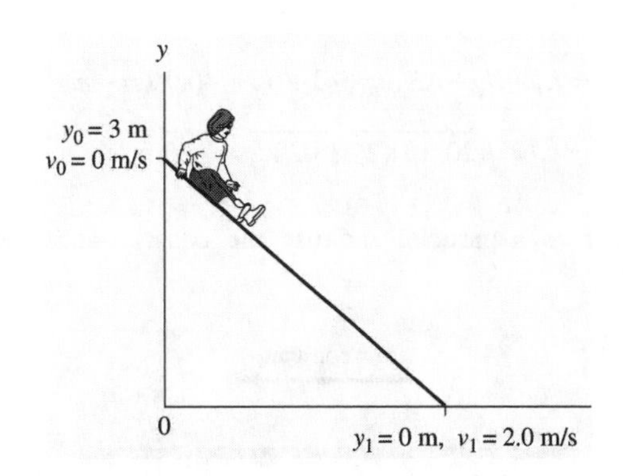


**Solve:** **(a)** $K_\text{i} = K_0 = \frac{1}{2}mv_0^2 = 0\text{ J} \quad U_\text{i} = U_{\text{g}0} = mgy_0 = (20\text{ kg})(9.8\text{ m/s}^2)(3.0\text{ m}) = 5.9\times10^2\text{ J}$

$$W_\text{ext} = 0\text{ J} \quad K_\text{f} = K_1 = \frac{1}{2}mv_1^2 = \frac{1}{2}(20\text{ kg})(2.0\text{ m/s})^2 = 40\text{ J} \quad U_\text{f} = U_{\text{g}1} = mgy_1 = 0\text{ J}$$

At the top of the slide, the child has gravitational potential energy of $5.9\times10^2$ J. This energy is transformed into thermal energy of the child's pants and the slide and the kinetic energy of the child. This energy transfer and transformation is shown on the energy bar chart.

**(b)**

Energy (J)

$K_\text{i}$ + $U_\text{i}$ + $W_\text{ext}$ = $K_\text{f}$ + $U_\text{f}$ + $\Delta E_\text{th}$

+
600
0
–

The change in the thermal energy of the slide and of the child's pants is $5.9\times10^2\text{ J} - 40\text{ J} = 5.5\times10^2\text{ J}$.

**11.29. Visualize:** The tension of 20.0 N in the cable is an external force that does work on the block $W_\text{ext} = (20.0\text{ N})(2.00\text{ m}) = 40.0\text{ J}$, increasing the gravitational potential energy of the block. We placed the origin of our coordinate system on the initial resting position of the block, so we have $U_\text{i} = 0\text{ J}$ and $U_\text{f} = mgy_\text{f} = (1.02\text{ kg})(9.8\text{ m/s}^2)(2.00\text{ m}) = 20.0\text{ J}$. Also, $K_\text{i} = 0\text{ J}$, and $\Delta E_\text{th} = 0\text{ J}$. The energy bar chart shows the energy transfers and transformations.

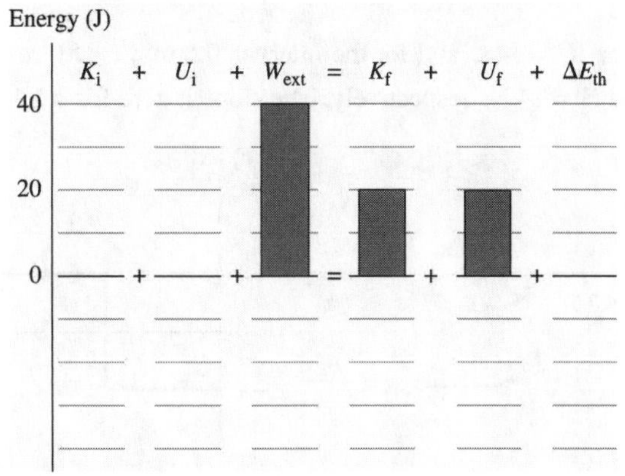

**Solve:** The conservation of energy equation is

$$K_i + U_i + W_{ext} = K_f + U_f + \Delta E_{th} \Rightarrow 0\ \text{J} + 0\ \text{J} + 40.0\ \text{J} = \frac{1}{2}mv_f^2 + 20.0\ \text{J} + 0\ \text{J}$$

$$\Rightarrow v_f = \sqrt{(20.0\ \text{J})(2)/1.02\ \text{kg}} = 6.26\ \text{m/s}$$

**11.35. Model:** Model the sprinter as a particle, and use the constant-acceleration kinematic equations and the definition of power in terms of velocity.
**Visualize:**

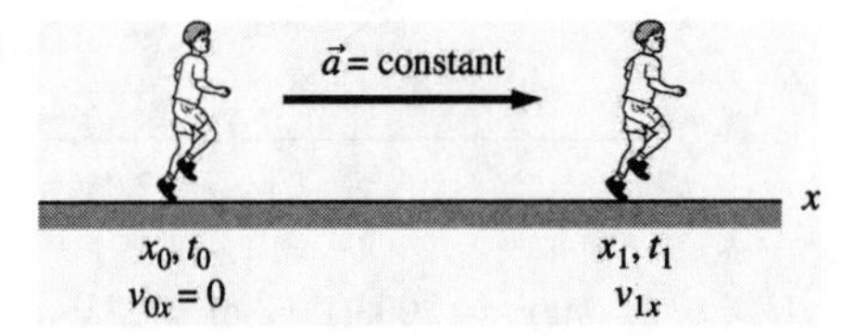


**Solve:** **(a)** We can find the acceleration from the kinematic equations and the horizontal force from Newton's second law. We have

$$x = x_0 + v_{0x}(t_1 - t_0) + \frac{1}{2}a_x(t_1 - t_0)^2 \Rightarrow 50\ \text{m} = 0\ \text{m} + 0\ \text{m} + \frac{1}{2}a_x(7.0\ \text{s} - 0\ \text{s})^2 \Rightarrow a_x = 2.04\ \text{m/s}^2$$

$$\Rightarrow F_x = ma_x = (50\ \text{kg})(2.04\ \text{m/s}^2) = 102\ \text{N}$$

**(b)** We obtain the sprinter's power output by using $P = \vec{F} \cdot \vec{v}$, where $\vec{v}$ is the sprinter's velocity. At $t = 2.0$ s the power is

$$P = (F_x)[v_{0x} + a_x(t - t_0)] = (102\ \text{N})[0\ \text{m/s} + (2.04\ \text{m/s}^2)(2.0\ \text{s} - 0\ \text{s})] = 0.42\ \text{kW}$$

The power at $t = 4.0$ s is 0.83 kW, and at $t = 6.0$ s the power is 1.25 kW.

**11.39. Model:** Use the relationship between a conservative force and potential energy.
**Visualize:** Please refer to Figure P11.39. We will obtain $U$ as a function of $x$ and $F_x$ as a function of $x$ by using the calculus techniques of integration and differentiation.
**Solve:** **(a)** For the interval $0\ \text{m} < x < 0.5\ \text{m}$, $F_x = (4x)$ N, where $x$ is in meters. This means

$$\frac{dU}{dx} = -F_x = -4x \Rightarrow U = -2x^2 + C_1 = -2x^2$$

where we have used $U = 0$ J at $x = 0$ m to obtain $C_1 = 0$. For the interval $0.5\ \text{m} < x < 1\ \text{m}$, $F_x = (-4x + 4)$ N. Likewise,

$$\frac{dU}{dx} = 4x - 4 \Rightarrow U = 2x^2 - 4x + C_2$$

Since $U$ should be continuous at the junction, we have the continuity condition

$$(-2x^2)_{x=0.5\ \text{m}} = (2x^2 - 4x + C_2)_{x=0.5\ \text{m}} \Rightarrow -0.5 = 0.5 - 2 + C_2 \Rightarrow C_2 = 1$$

$U$ remains constant for $x \geq 1$ m.
**(b)** For the interval $0\ \text{m} < x < 0.5\ \text{m}$, $U = +4x$, and for the interval $0.5\ \text{m} < x < 1.0\ \text{m}$, $U = -4x + 4$, where $x$ is in meters. The derivatives give $F_x = -4$ N and $F_x = +4$ N, respectively. The slope is zero for $x \geq 1$ m.

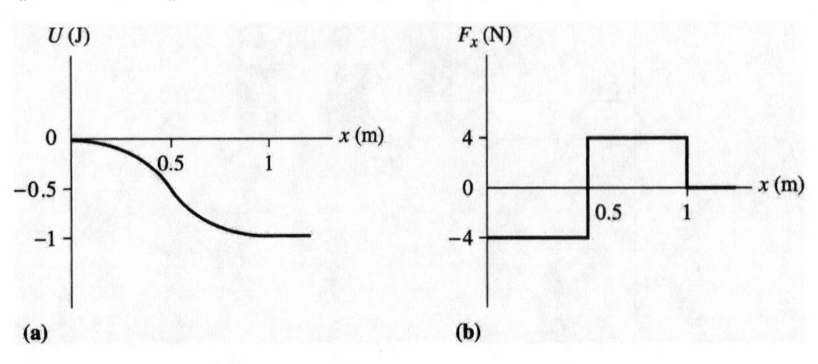

**11.41. Model:** Model the elevator as a particle.
**Visualize:**

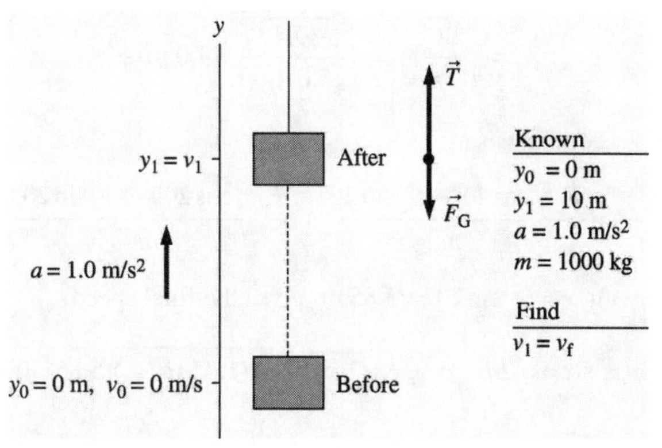


**Solve:** **(a)** The work done by gravity on the elevator is

$$W_g = -\Delta U_g = mgy_0 - mgy_1 = -mg(y_1 - y_0) = -(1000\text{ kg})(9.8\text{ m/s}^2)(10\text{ m}) = -9.8\times10^4\text{ J}$$

**(b)** The work done by the tension in the cable on the elevator is

$$W_T = T(\Delta y)\cos 0° = T(y_1 - y_0) = T(10\text{ m})$$

To find $T$ we write Newton's second law for the elevator:

$$\sum F_y = T - F_G = ma_y \Rightarrow T = F_G + ma_y = m(g + a_y) = (1000\text{ kg})(9.8\text{ m/s}^2 + 1.0\text{ m/s}^2)$$
$$= 1.08\times10^4\text{ N} \Rightarrow W_T = (1.08\times10^4\text{ N})(10\text{ m}) = 1.08\times10^5\text{ J}$$

**(c)** The work-kinetic energy theorem is

$$W_{net} = W_g + W_T = \Delta K = K_f - K_i = K_f - \frac{1}{2}mv_0^2 \Rightarrow K_f = W_g + W_T + \frac{1}{2}mv_0^2$$
$$\Rightarrow K_f = (-9.8\times10^4\text{ J}) + (1.08\times10^5\text{ J}) + \frac{1}{2}(1000\text{ kg})(0\text{ m/s})^2 = 1.0\times10^4\text{ J}$$

**(d)** $K_f = \frac{1}{2}mv_f^2 \Rightarrow 1.0\times10^4\text{ J} = \frac{1}{2}(1000\text{ kg})v_f^2 \Rightarrow v_f = 4.5\text{ m/s}$

**11.43. Model:** Model the crate as a particle, and use the work-kinetic energy theorem.
**Visualize:**

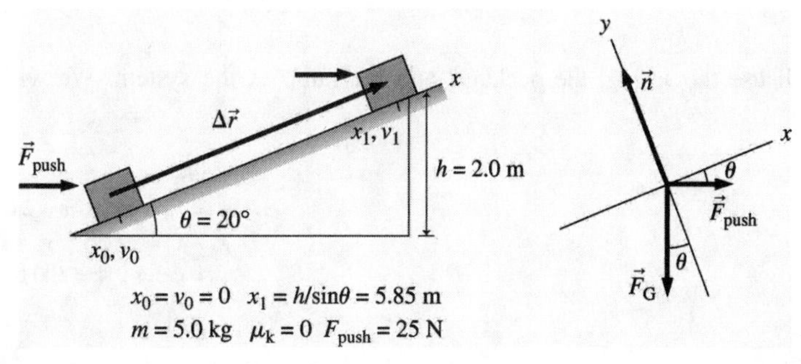


**Solve:** **(a)** The work-kinetic energy theorem is $\Delta K = \frac{1}{2}mv_1^2 - \frac{1}{2}mv_0^2 = \frac{1}{2}mv_1^2 = W_{total}$. Three forces act on the box, so $W_{total} = W_{grav} + W_n + W_{push}$. The normal force is perpendicular to the motion, so $W_n = 0$ J. The other two forces do the following amount of work:

$$W_{push} = \vec{F}_{push}\cdot\Delta\vec{r} = F_{push}\Delta x\cos 20° = 137.4\text{ J} \qquad W_{grav} = \vec{F}_G\cdot\Delta\vec{r} = (F_G)_x\Delta x = (-mg\sin 20°)\Delta x = -98.0\text{ J}$$

Thus, $W_{\text{total}} = 39.4\text{ J}$, leading to a speed at the top of the ramp equal to

$$v_1 = \sqrt{\frac{2W_{\text{total}}}{m}} = \sqrt{\frac{2(39.4\text{ J})}{5.0\text{ kg}}} = 4.0\text{ m/s}$$

**(b)** The $x$-component of Newton's second law is

$$a_x = a = \frac{(F_{\text{net}})_x}{m} = \frac{F_{\text{push}}\cos 20° - F_G \sin 20°}{m} = \frac{F_{\text{push}}\cos 20° - mg\sin 20°}{m} = 1.35\text{ m/s}^2$$

Constant-acceleration kinematics with $x_1 = h/\sin 20° = 5.85\text{ m}$ gives the final speed

$$v_1^2 = v_0^2 + 2a(x_1 - x_0) = 2ax_1 \Rightarrow v_1 = \sqrt{2ax_1} = \sqrt{2(1.35\text{ m/s}^2)(5.85\text{ m})} = 4.0\text{ m/s}$$

**11.47. Model:** Model the suitcase as a particle, use the model of kinetic friction, and use the work-kinetic energy theorem.
**Visualize:**

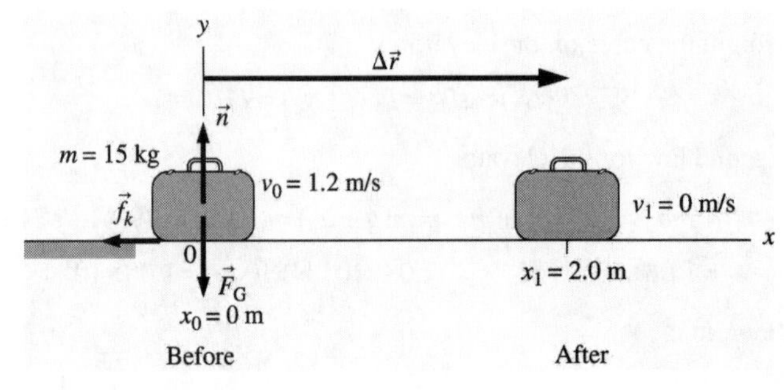


The net force on the suitcase is $\vec{F}_{\text{net}} = \vec{f}_k$.
**Solve:** The work-kinetic energy theorem is

$$W_{\text{net}} = \Delta K = \frac{1}{2}mv_1^2 - \frac{1}{2}mv_0^2 \Rightarrow \vec{F}_{\text{net}} \cdot \Delta\vec{r} = \vec{f}_k \cdot \Delta\vec{r} = 0\text{ J} - \frac{1}{2}mv_0^2 \Rightarrow (f_k)(x_1 - x_0)\cos 180° = -\frac{1}{2}mv_0^2$$

$$\Rightarrow -\mu_k mg(x_1 - x_0) = -\frac{1}{2}mv_0^2 \Rightarrow \mu_k = \frac{v_0^2}{2g(x_1 - x_0)} = \frac{(1.2\text{ m/s})^2}{2(9.8\text{ m/s}^2)(2.0\text{ m} - 0\text{ m})} = 0.037$$

**Assess:** Friction transforms kinetic energy of the suitcase into thermal energy. In response, the suitcase slows down and comes to rest.

**11.49. Model:** We will use the spring, the package, and the ramp as the system. We will model the package as a particle.
**Visualize:**

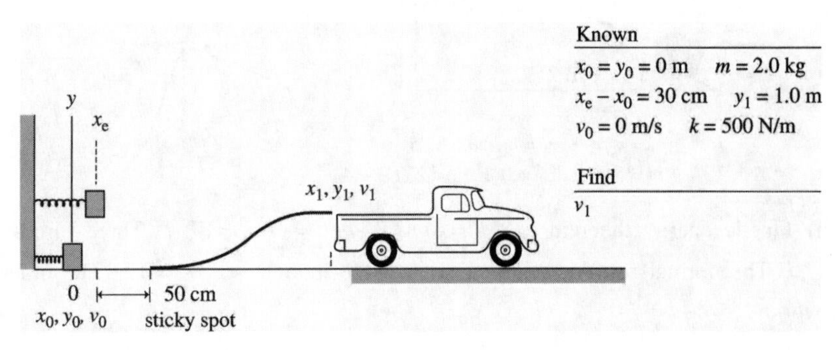


We place the origin of our coordinate system on the end of the spring when it is compressed and is in contact with the package to be shot.

**Model:** **(a)** The energy conservation equation is

$$K_1 + U_{g1} + U_{s1} + \Delta E_{th} = K_0 + U_{g0} + U_{s0} + W_{ext}$$

$$\frac{1}{2}mv_1^2 + mgy_1 + \frac{1}{2}k(x_e - x_e)^2 + \Delta E_{th} = \frac{1}{2}mv_0^2 + mgy_0 + \frac{1}{2}k(\Delta x)^2 + W_{ext}$$

Using $y_1 = 1$ m, $\Delta E_{th} = 0$ J (note the frictionless ramp), $v_0 = 0$ m/s, $y_0 = 0$ m, $\Delta x = 30$ cm, and $W_{ext} = 0$ J, we get

$$\frac{1}{2}mv_1^2 + mg(1\text{ m}) + 0\text{ J} + 0\text{ J} = 0\text{ J} + 0\text{ J} + \frac{1}{2}k(0.30\text{ m})^2 + 0\text{ J}$$

$$\frac{1}{2}(2.0\text{ kg})v_1^2 + (2.0\text{ kg})(9.8\text{ m/s}^2)(1\text{ m}) = \frac{1}{2}(500\text{ N/m})(0.30\text{ m})^2$$

$$\Rightarrow v_1 = 1.70\text{ m/s}$$

**(b)** How high can the package go after crossing the sticky spot? If the package can reach $y_1 \geq 1.0$ m before stopping ($v_1 = 0$), then it makes it. But if $y_1 < 1.0$ m when $v_1 = 0$, it does not. The friction of the sticky spot generates thermal energy

$$\Delta E_{th} = (\mu_k mg)\Delta x = (0.30)(2.0\text{ kg})(9.8\text{ m/s}^2)(0.50\text{ m}) = 2.94\text{ J}$$

The energy conservation equation is now

$$\tfrac{1}{2}mv_1^2 + mgy_1 + \Delta E_{th} = \tfrac{1}{2}k(\Delta x)^2$$

If we set $v_1 = 0$ m/s to find the highest point the package can reach, we get

$$y_1 = \left(\tfrac{1}{2}k(\Delta x)^2 - \Delta E_{th}\right)/mg = \left(\tfrac{1}{2}(500\text{ N/m})(0.30\text{ m})^2 - 2.94\text{ J}\right)/(2.0\text{ kg})(9.8\text{ m/s}^2) = 0.998\text{ m}$$

The package does not make it. It just barely misses.

**11.53. Model:** Use the particle model for the ice skater, the model of kinetic/static friction, and the work-kinetic energy theorem.

**Visualize:**

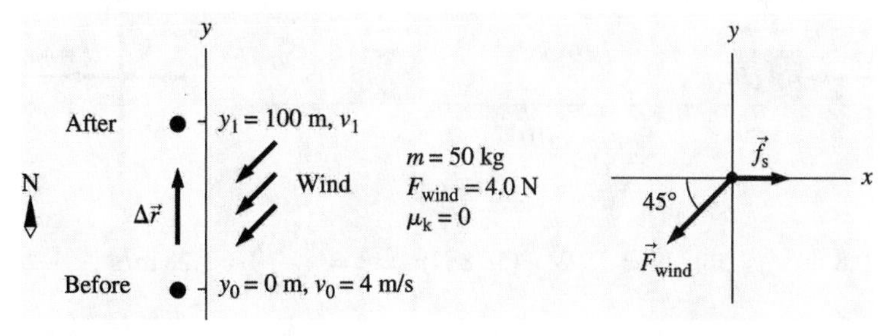


**Solve:** **(a)** The work-kinetic energy theorem is

$$\Delta K = \frac{1}{2}mv_1^2 - \frac{1}{2}mv_0^2 = W_{net} = W_{wind}$$

There is no kinetic friction along her direction of motion. Static friction acts to prevent her skates from slipping sideways on the ice, but this force is perpendicular to the motion and does not contribute to a change in thermal energy. The angle between $\vec{F}_{wind}$ and $\Delta\vec{r}$ is $\theta = 135°$, so

$$W_{wind} = \vec{F}_{wind} \cdot \Delta\vec{r} = F_{wind}\Delta y\cos 135° = (4\text{ N})(100\text{ m})\cos 135° = -282.8\text{ J}$$

Thus, her final speed is

$$v_1 = \sqrt{v_0^2 + \frac{2W_{wind}}{m}} = 2.16\text{ m/s}$$

**(b)** If the skates don't slip, she has no acceleration in the $x$-direction and so $(F_{net})_x = 0$ N. That is:

$$f_s - F_{wind}\cos 45° = 0\text{ N} \Rightarrow f_s = F_{wind}\cos 45° = 2.83\text{ N}$$

Now there is an upper limit to the static friction: $f_s \le (f_s)_{max} = \mu_s mg$. To not slip requires

$$\mu_s \ge \frac{f_s}{mg} = \frac{2.83 \text{ N}}{(50 \text{ kg})(9.8 \text{ m/s}^2)} = 0.0058$$

Thus, the minimum value of $\mu_s$ is 0.0058.

**Assess:** The work done by the wind on the ice skater is negative, because the wind slows the skater down.

**11.67. Solve:** The net force on a car moving at a steady speed is zero. The motion is opposed both by rolling friction and by air resistance. Thus the propulsion force provided by the drive wheels must be $F_{car} = \mu_r mg + \frac{1}{4}Av^2$, where $\mu_r$ is the rolling friction, *m* is the mass, *A* is the cross-section area, and *v* is the car's velocity. The power required to move the car at speed *v* is

$$P = F_{car}v = \mu_r mgv + \frac{1}{4}Av^3$$

Since the maximum power output is 200 hp and 75% of the power reaches the drive wheels, $P = (200 \text{ hp})(0.75) = 150 \text{ hp}$. Thus,

$$(150 \text{ hp})\left(\frac{746 \text{ W}}{1 \text{ hp}}\right) = (0.02)(1500 \text{ kg})(9.8 \text{ m/s}^2)v + \frac{1}{4}(1.6 \text{ m})(1.4 \text{ m})v^3$$

$$\Rightarrow 0.56\,v^3 + 294\,v - 111{,}900 = 0 \Rightarrow v = 55.5 \text{ m/s}$$

The easiest way to solve this equation is through iterations by trial and error.

**Assess:** A speed of $55.5 \text{ m/s} \approx 110 \text{ mph}$ is very reasonable.

**11.71. (a)** If you expend 75 W of power to push a 30 kg sled on a surface where the coefficient of kinetic friction between the sled and the surface is $\mu_k = 0.20$, what speed will you be able to maintain?

**(b)**

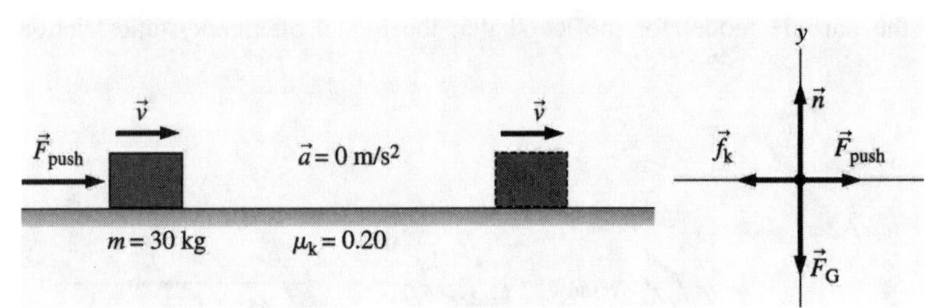


**(c)** $F_{push} = (0.20)(30 \text{ kg})(9.8 \text{ m/s}^2) = 58.8 \text{ N} \Rightarrow 75 \text{ W} = (58.8 \text{ N})v \Rightarrow v = \frac{75 \text{ W}}{58.8 \text{ N}} = 1.28 \text{ m/s}$

# 12

# ROTATION OF A RIGID BODY

**12.1. Model:** A spinning skater, whose arms are outstretched, is a rigid rotating body.
**Visualize:**

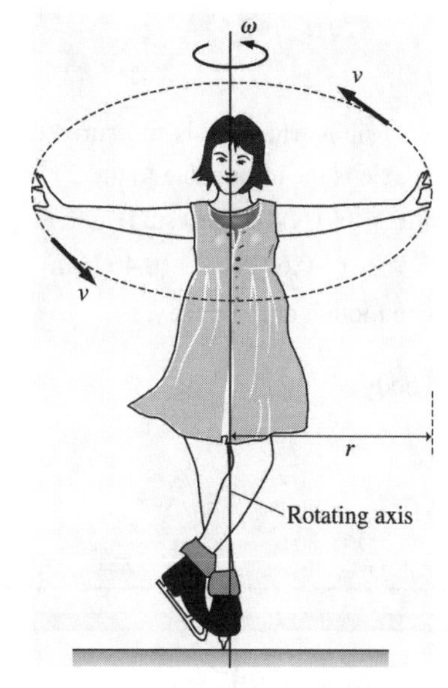


**Solve:** The speed $v = r\omega$, where $r = 140\text{ cm}/2 = 0.70\text{ m}$. Also, $180\text{ rpm} = (180)2\pi/60\text{ rad/s} = 6\pi\text{ rad/s}$. Thus, $v = (0.70\text{ m})(6\pi\text{ rad/s}) = 13.2\text{ m/s}$.
**Assess:** A speed of $13.2\text{ m/s} \approx 26\text{ mph}$ for the hands is a little high, but reasonable.

**12.7. Visualize:** Please refer to Figure EX12.7. The coordinates of the three masses $m_A$, $m_B$, and $m_C$ are (0 cm, 10 cm), (10 cm, 10 cm), and (10 cm, 0 cm), respectively.
**Solve:** The coordinates of the center of mass are

$$x_{cm} = \frac{m_A x_A + m_B x_B + m_C x_C}{m_A + m_B + m_C} = \frac{(100\text{ g})(0\text{ cm}) + (200\text{ g})(10\text{ cm}) + (300\text{ g})(10\text{ cm})}{(100\text{ g} + 200\text{ g} + 300\text{ g})} = 8.3\text{ cm}$$

$$y_{cm} = \frac{m_A y_A + m_B y_B + m_C y_C}{m_A + m_B + m_C} = \frac{(100\text{ g})(10\text{ cm}) + (200\text{ g})(10\text{ cm}) + (300\text{ g})(0\text{ cm})}{(100\text{ g} + 200\text{ g} + 300\text{ g})} = 5.0\text{ cm}$$

**12.17. Model:** The door is a slab of uniform density.
**Solve:** (a) The hinges are at the edge of the door, so from Table 12.2,

$$I = \frac{1}{3}(25\text{ kg})(0.91\text{ m})^2 = 6.9\text{ kg m}^2$$

**(b)** The distance from the axis through the center of mass along the height of the door is $d = \left(\frac{0.91 \text{ cm}}{2} - 0.15 \text{ cm}\right) = 0.305$ cm. Using the parallel–axis theorem,

$$I = I_{cm} + Md^2 = \frac{1}{12}(25 \text{ kg})(0.91 \text{ m})^2 + (25 \text{ kg})(0.305 \text{ cm})^2 = 4.1 \text{ kg m}^2$$

**Assess:** The moment of inertia is less for a parallel axis through a point closer to the center of mass.

**12.19. Visualize:**

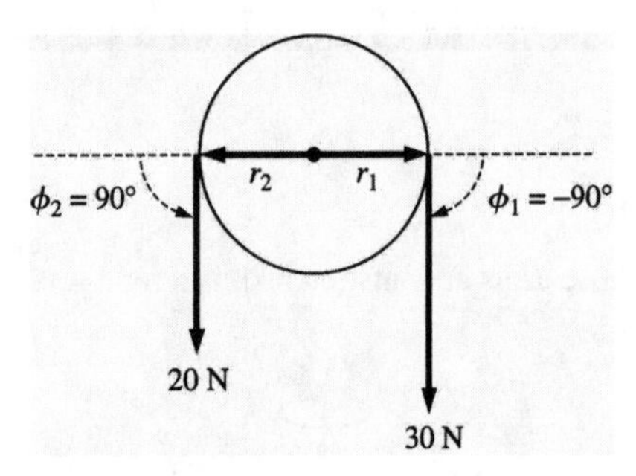


**Solve:** Torque by a force is defined as $\tau = Fr\sin\phi$ where $\phi$ is measured counterclockwise from the $\vec{r}$ vector to the $\vec{F}$ vector. The net torque on the pulley about the axle is the torque due to the 30 N force plus the torque due to the 20 N force:

$$\begin{aligned}(30 \text{ N})r_1 \sin\phi_1 + (20 \text{ N})r_2 \sin\phi_2 &= (30 \text{ N})(0.02 \text{ m}) \sin(-90°) + (20 \text{ N})(0.02 \text{ m}) \sin(90°) \\ &= (-0.60 \text{ N m}) + (0.40 \text{ N m}) = -0.20 \text{ N m}\end{aligned}$$

**Assess:** A negative torque causes a clockwise motion of the pulley.

**12.23. Model:** The beam is a solid rigid body.
**Visualize:**

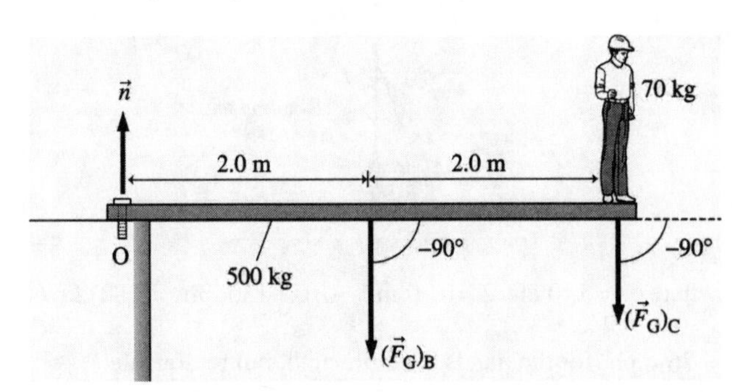


The steel beam experiences a torque due to the gravitational force on the construction worker $(\vec{F}_G)_C$ and the gravitational force on the beam $(\vec{F}_G)_B$. The normal force exerts no torque since the net torque is calculated about the point where the beam is bolted into place.

**Solve:** The net torque on the steel beam about point O is the sum of the torque due to $(\vec{F}_G)_C$ and the torque due to $(\vec{F}_G)_B$. The gravitational force on the beam acts at the center of mass.

$$\begin{aligned}\tau &= ((F_G)_C)(4.0 \text{ m})\sin(-90°) + ((F_G)_B)(2.0 \text{ m})\sin(-90°) \\ &= -(70 \text{ kg})(9.80 \text{ m/s}^2)(4.0 \text{ m}) - (500 \text{ kg})(9.80 \text{ m/s}^2)(2.0 \text{ m}) = -12.5 \text{ kN m}\end{aligned}$$

The negative torque means these forces would cause the beam to rotate clockwise. The magnitude of the torque is 12.5 kN m.

**12.27. Model:** Two balls connected by a rigid, massless rod are a rigid body rotating about an axis through the center of mass. Assume that the size of the balls is small compared to 1 m.

**Visualize:**

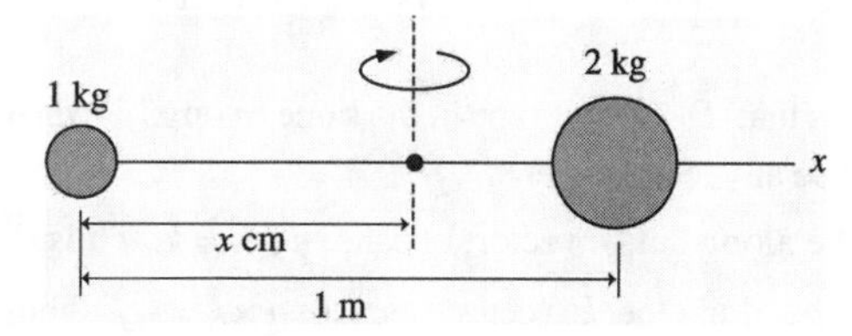


We placed the origin of the coordinate system on the 1.0 kg ball.
**Solve:** The center of mass and the moment of inertia are

$$x_{cm} = \frac{(1.0 \text{ kg})(0 \text{ m}) + (2.0 \text{ kg})(1.0 \text{ m})}{(1.0 \text{ kg} + 2.0 \text{ kg})} = 0.667 \text{ m and } y_{cm} = 0 \text{ m}$$

$$I_{\text{about cm}} = \sum m_i r_i^2 = (1.0 \text{ kg})(0.667 \text{ m})^2 + (2.0 \text{ kg})(0.333 \text{ m})^2 = 0.667 \text{ kg m}^2$$

We have $\omega_f = 0$ rad/s, $t_f - t_i = 5.0$ s, and $\omega_i = -20 \text{ rpm} = -20(2\pi \text{ rad}/60 \text{ s}) = -\frac{2}{3}\pi \text{rad/s}$, so $\omega_f = \omega_i + \alpha(t_f - t_i)$ becomes

$$0 \text{ rad/s} = \left(-\frac{2\pi}{3}\text{rad/s}\right) + \alpha(5.0 \text{ s}) \Rightarrow \alpha = \frac{2\pi}{15}\text{rad/s}^2$$

Having found $I$ and $\alpha$, we can now find the torque $\tau$ that will bring the balls to a halt in 5.0 s:

$$\tau = I_{\text{about cm}}\alpha = \left(\frac{2}{3}\text{kg m}^2\right)\left(\frac{2\pi}{15} \text{ rad/s}^2\right) = \frac{4\pi}{45} \text{ N m} = 0.28 \text{ N m}$$

The magnitude of the torque is 0.28 N m, applied in the counterclockwise direction.

**12.31. Model:** The rod is in rotational equilibrium, which means that $\tau_{net} = 0$.
**Visualize:**

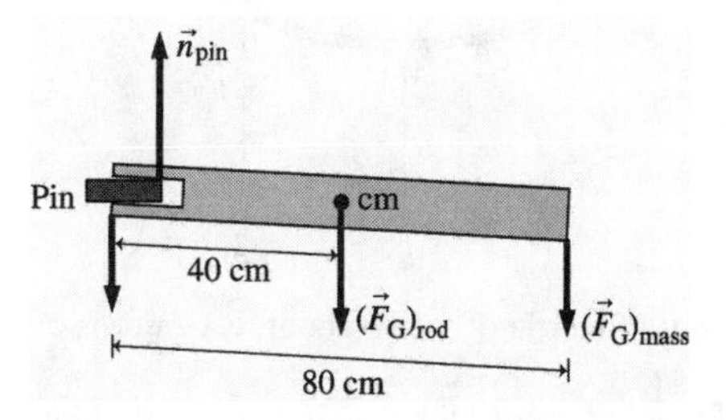


As the gravitational force on the rod and the hanging mass pull down (the rotation of the rod is exaggerated in the figure), the rod touches the pin at two points. The piece of the pin at the very end pushes down on the rod; the right end of the pin pushes up on the rod. To understand this, hold a pen or pencil between your thumb and forefinger, with your thumb on top (pushing down) and your forefinger underneath (pushing up).
**Solve:** Calculate the torque about the left end of the rod. The downward force exerted by the pin acts through this point, so it exerts no torque. To prevent rotation, the pin's normal force $\vec{n}_{pin}$ exerts a positive torque (ccw about the left end) to balance the negative torques (cw) of the gravitational force on the mass and rod. The gravitational force on the rod acts at the center of mass, so

$$\tau_{net} = 0 \text{ N m} = \tau_{pin} - (0.40 \text{ m})(2.0 \text{ kg})(9.8 \text{ m/s}^2) - (0.80 \text{ m})(0.50 \text{ kg})(9.8 \text{ m/s}^2)$$

$$\Rightarrow \tau_{pin} = 11.8 \text{ N m}$$

**12.35. Solve:** **(a)** According to Equation 12.35, the speed of the center of mass of the tire is

$$v_{cm} = R\omega = 20 \text{ m/s} \Rightarrow \omega = \frac{v_{cm}}{R} = \frac{20 \text{ m/s}}{0.30 \text{ m}} = 66.67 \text{ rad/s} = (66.7)\left(\frac{60}{2\pi}\right) \text{ rpm} = 6.4\times10^2 \text{ rpm}$$

**(b)** The speed at the top edge of the tire relative to the ground is $v_{top} = 2v_{cm} = 2(20 \text{ m/s}) = 40$ m/s.

**(c)** The speed at the bottom edge of the tire relative to ground is $v_{\text{bottom}} = 0$ m/s.

**12.43. Solve:** **(a)** $\vec{C} \times \vec{D} = 0$ implies that $\vec{D}$ must also be in the same or opposite direction as the $\vec{C}$ vector or zero, because $\hat{i} \times \hat{i} = 0$. Thus $\vec{D} = n\hat{i}$, where $n$ could be any real number.
**(b)** $\vec{C} \times \vec{E} = 6\hat{k}$ implies that $\vec{E}$ must be along the $\hat{j}$ vector, because $\hat{i} \times \hat{j} = \hat{k}$. Thus $\vec{E} = 2\hat{j}$.
**(c)** $\vec{C} \times \vec{F} = -3\hat{j}$ implies that $\vec{F}$ must be along the $\hat{k}$ vector, because $\hat{i} \times \hat{k} = -\hat{j}$. Thus $\vec{F} = 1\hat{k}$.

**12.49. Model:** The disk is a rotating rigid body.
**Visualize:** Please refer to Figure EX12.49.
**Solve:** From Table 12.2, the moment of inertial of the disk about its center is

$$I = \frac{1}{2}MR^2 = \frac{1}{2}(2.0 \text{ kg})(0.020 \text{ m})^2 = 4.0 \times 10^{-4} \text{ kg m}^2$$

The angular velocity $\omega$ is 600 rpm $= 600 \times 2\pi/60$ rad/s $= 20\pi$ rad/s. Thus, $L = I\omega = (4.0 \times 10^{-4} \text{ kg m}^2)(20\pi \text{ rad/s}) =$ 0.025 kg m$^2$/s. If we wrap our right fingers in the direction of the disk's rotation, our thumb will point in the $-x$ direction. Consequently,

$$\vec{L} = -0.025\,\hat{i} \text{ kg m}^2\text{/s} = (0.025 \text{ kg m}^2\text{/s, into page})$$

**12.55. Model:** The disk is a rigid rotating body. The axis is perpendicular to the plane of the disk.
**Visualize:**

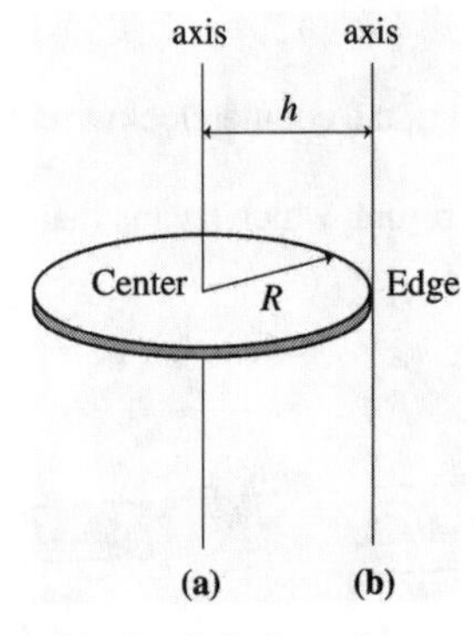


**Solve:** **(a)** From Table 12.2, the moment of inertia of a disk about its center is

$$I = \frac{1}{2}MR^2 = \frac{1}{2}(2.0 \text{ kg})(0.10 \text{ m})^2 = 0.010 \text{ kg m}^2$$

**(b)** To find the moment of inertia of the disk through the edge, we can make use of the parallel axis theorem:

$$I = I_{\text{center}} + Mh^2 = (0.010 \text{ kg m}^2) + (2.0 \text{ kg})(0.10 \text{ m})^2 = 0.030 \text{ kg m}^2$$

**Assess:** The larger moment of inertia about the edge means there is more inertia to rotational motion about the edge than about the center.

**12.59. Model:** The plate has uniform density.
**Visualize:**

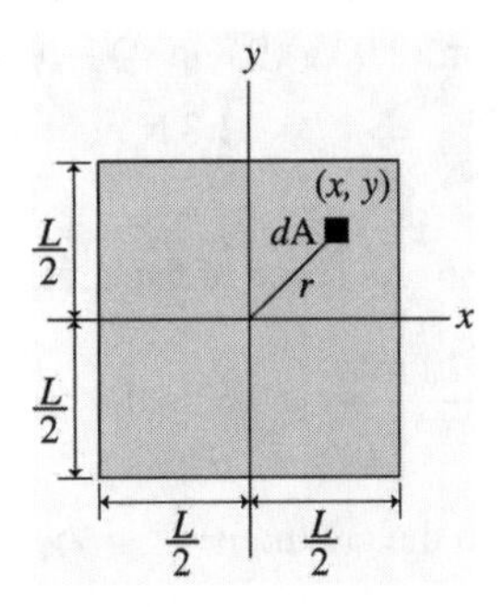

**Solve:** The moment of inertia is

$$I = \int r^2 dm.$$

Let the mass of the plate be $M$. Its area is $L^2$. A region of area $dA$ located at $(x,y)$ has mass $dm = \frac{M}{A}dA = \frac{M}{L^2}dx\,dy$. The distance from the axis of rotation to the point $(x, y)$ is $r = \sqrt{x^2 + y^2}$. With $-\frac{L}{2} \le x \le \frac{L}{2}$ and $-\frac{L}{2} \le y \le \frac{L}{2}$,

$$I = \int_{-\frac{L}{2}}^{\frac{L}{2}} \int_{-\frac{L}{2}}^{\frac{L}{2}} (x^2 + y^2)\left(\frac{M}{L^2}\right) dx\,dy = \frac{M}{L^2} \int_{-\frac{L}{2}}^{\frac{L}{2}} \left(\frac{x^3}{3} + y^2 x\right)\Bigg|_{-\frac{L}{2}}^{\frac{L}{2}} dy$$

$$= \frac{M}{L^2} \int_{-\frac{L}{2}}^{\frac{L}{2}} \left(\frac{L^3}{24} + \frac{y^2 L}{2} - \left(\frac{-L^3}{24} - \frac{y^2 L}{2}\right)\right) dy = \frac{M}{L^2} \int_{-\frac{L}{2}}^{\frac{L}{2}} \left(\frac{L^3}{12} + Ly^2\right) dy = \frac{M}{L^2}\left(\frac{L^4}{12} + L\frac{y^3}{3}\Bigg|_{-\frac{L}{2}}^{\frac{L}{2}}\right)$$

$$= \frac{M}{L^2}\left(\frac{L^4}{12} + L\left(\frac{L^3}{24} + \frac{L^3}{24}\right)\right) = \frac{1}{6}ML^2$$

**12.63. Model:** The structure is a rigid body.
**Visualize:**

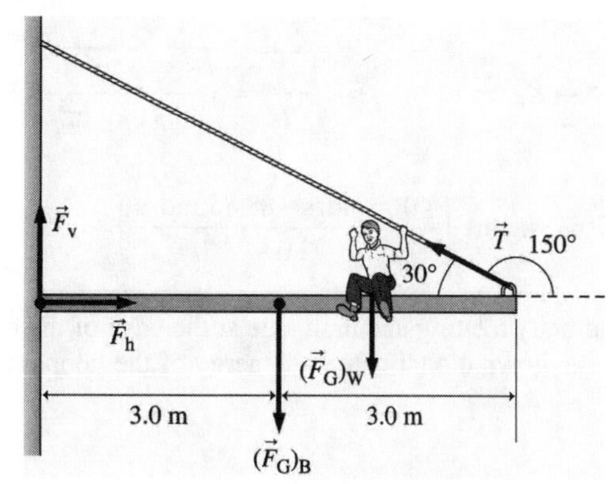


**Solve:** We pick the left end of the beam as our pivot point. We don't need to know the forces $F_h$ and $F_v$ because the pivot point passes through the line of application of $F_h$ and $F_v$ and therefore these forces do not exert a torque. For the beam to stay in equilibrium, the net torque about this point is zero. We can write

$$\tau_{\text{about left end}} = -(F_G)_B(3.0 \text{ m}) - (F_G)_W(4.0 \text{ m}) + (T \sin 150°)(6.0 \text{ m}) = 0 \text{ N m}$$

Using $(F_G)_B = (1450 \text{ kg})(9.8 \text{ m/s}^2)$ and $(F_G)_W = (80 \text{ kg})(9.8 \text{ m/s}^2)$, the torque equation can be solved to yield $T = 15{,}300$ N. The tension in the cable is slightly more than the cable rating. The worker should be worried.

**12.69. Model:** The flywheel is a rigid body rotating about its central axis.
**Visualize:**

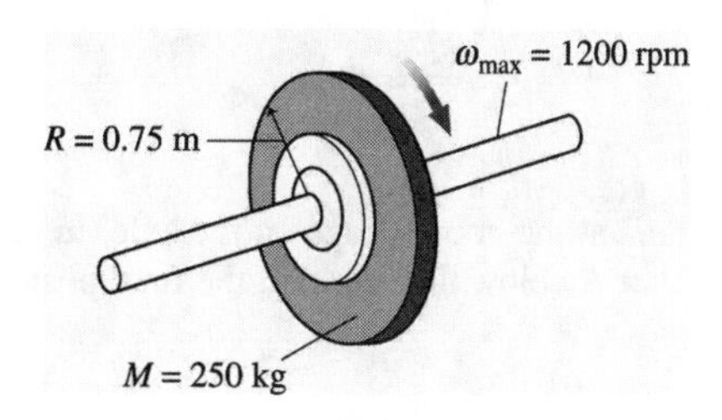

**Solve:** **(a)** The radius of the flywheel is $R = 0.75$ m and its mass is $M = 250$ kg. The moment of inertia about the axis of rotation is that of a disk:

$$I = \frac{1}{2}MR^2 = \frac{1}{2}(250 \text{ kg})(0.75 \text{ m})^2 = 70.31 \text{ kg m}^2$$

The angular acceleration is calculated as follows:

$$\tau_{\text{net}} = I\alpha \Rightarrow \alpha = \tau_{\text{net}}/I = (50 \text{ N m})/(70.31 \text{ kg m}^2) = 0.711 \text{ rad/s}^2$$

Using the kinematic equation for angular velocity gives

$$\omega_1 = \omega_0 + \alpha(t_1 - t_0) = 1200 \text{ rpm} = 40\ \pi \text{ rad/s} = 0 \text{ rad/s} + 0.711 \text{ rad/s}^2(t_1 - 0 \text{ s})$$
$$\Rightarrow t_1 = 177 \text{ s}$$

**(b)** The energy stored in the flywheel is rotational kinetic energy:

$$K_{\text{rot}} = \frac{1}{2}I\omega_1^2 = \frac{1}{2}(70.31 \text{ kg m}^2)(40\pi \text{ rad/s})^2 = 5.55\times10^5 \text{ J}$$

The energy stored is $5.6\times10^5$ J.

**(c)** Average power delivered $= \dfrac{\text{energy delivered}}{\text{time interval}} = \dfrac{(5.55\times10^5 \text{ J})/2}{2.0 \text{ s}} = 1.39\times10^5 \text{ W} = 139 \text{ kW}$

**(d)** Because $\tau = I\alpha,\ \Rightarrow \tau = I\dfrac{\Delta\omega}{\Delta t} = I\left(\dfrac{\omega_{\text{full energy}} - \omega_{\text{half energy}}}{\Delta t}\right)$. $\omega_{\text{full energy}} = \omega_1$ (from part (a)) $= 40\pi$ rad/s. $\omega_{\text{half energy}}$ can be obtained as:

$$\frac{1}{2}I\omega_{\text{half energy}}^2 = \frac{1}{2}K_{\text{rot}} \Rightarrow \omega_{\text{half energy}} = \sqrt{\frac{K_{\text{rot}}}{I}} = \sqrt{\frac{5.55\times10^5 \text{ J}}{70.31 \text{ kg m}^2}} = 88.85 \text{ rad/s}$$

Thus

$$\tau = (70.31 \text{ kg m}^2)\left(\frac{40\ \pi \text{ rad/s} - 88.85 \text{ rad/s}}{2.0 \text{ s}}\right) = 1.30 \text{ kN m}$$

**12.77. Model:** The hoop is a rigid body rotating about an axle at the edge of the hoop. The gravitational torque on the hoop causes it to rotate, transforming the gravitational potential energy of the hoop's center of mass into rotational kinetic energy.

**Visualize:**

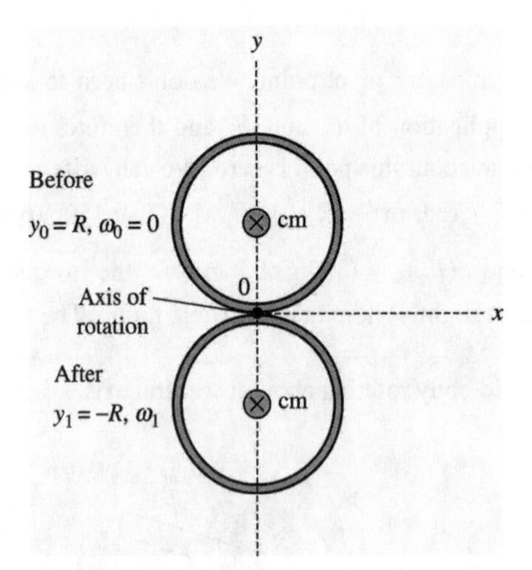


We placed the origin of the coordinate system at the hoop's edge on the axle. In the initial position, the center of mass is a distance $R$ above the origin, but it is a distance $R$ below the origin in the final position.

**Solve:** (**a**) Applying the parallel-axis theorem, $I_{\text{edge}} = I_{\text{cm}} + mR^2 = mR^2 + mR^2 = 2mR^2$. Using this expression in the energy conservation equation $K_f + U_{gf} = K_i + U_{gi}$ yields:

$$\frac{1}{2}I_{\text{edge}}\omega_1^2 + mgy_1 = \frac{1}{2}I_{\text{edge}}\omega_0^2 + mgy_0 \quad \frac{1}{2}(2mR^2)\omega_1^2 - mgR = 0\text{ J} + mgR \Rightarrow \omega_1 = \sqrt{\frac{2g}{R}}$$

(**b**) The speed of the lowest point on the hoop is

$$v = (\omega_1)(2R) = \sqrt{\frac{2g}{R}}(2R) = \sqrt{8gR}$$

**Assess:** Note that the speed of the lowest point on the loop involves a distance of $2R$ instead of $R$.

**12.85. Model:** The mechanical energy of both the hoop ($h$) and the sphere ($s$) is conserved. The initial gravitational potential energy is transformed into kinetic energy as the objects roll down the slope. The kinetic energy is a combination of translational and rotational kinetic energy. We also assume no slipping of the hoop or of the sphere.
**Visualize:**

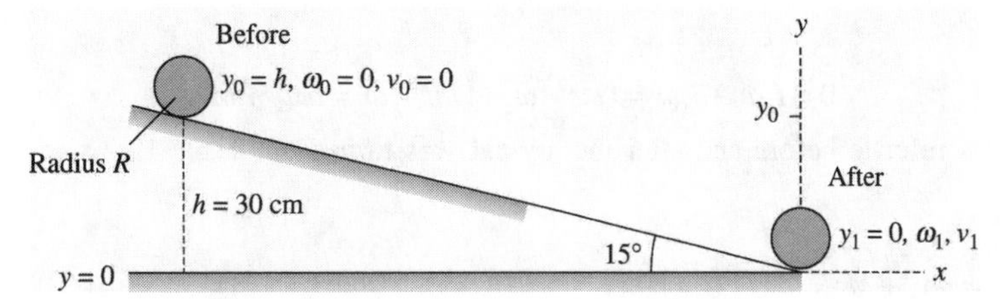


The zero of gravitational potential energy is chosen at the bottom of the slope.
**Solve:** (**a**) The energy conservation equation for the sphere or hoop $K_f + U_{gf} = K_i + U_{gi}$ is

$$\frac{1}{2}I(\omega_1)^2 + \frac{1}{2}m(v_1)^2 + mgy_1 = \frac{1}{2}I(\omega_0)^2 + \frac{1}{2}m(v_0)^2 + mgy_0$$

For the sphere, this becomes

$$\frac{1}{2}\left(\frac{2}{5}mR^2\right)\frac{(v_1)_s^2}{R^2} + \frac{1}{2}m(v_1)_s^2 + 0\text{ J} = 0\text{ J} + 0\text{ J} + mgh_s$$

$$\Rightarrow \frac{7}{10}(v_1)_s^2 = gh \Rightarrow (v_1)_s = \sqrt{10gh/7} = \sqrt{10(9.8\text{ m/s}^2)(0.30\text{ m})/7} = 2.05\text{ m/s}$$

For the hoop, this becomes

$$\frac{1}{2}(mR^2)\frac{(v_1)_h^2}{R^2} + \frac{1}{2}m(v_1)_h^2 + 0\text{ J} = 0\text{ J} + 0\text{ J} + mgh_{\text{hoop}}$$

$$\Rightarrow h_{\text{hoop}} = \frac{(v_1)_h^2}{g}$$

For the hoop to have the same velocity as that of the sphere,

$$h_{\text{hoop}} = \frac{(v_1)_s^2}{g} = \frac{(2.05\text{ m/s})^2}{9.8\text{ m/s}^2} = 42.9\text{ cm}$$

The hoop should be released from a height of 43 cm.
(**b**) As we see in part (a), the speed of a hoop at the bottom depends only on the starting height and not on the mass or radius. So the answer is No.

**12.89. Model:** The toy car is a particle located at the rim of the track. The track is a cylindrical hoop rotating about its center, which is an axis of symmetry. No net torques are present on the track, so the angular momentum of the car and track is conserved.

**Visualize:**

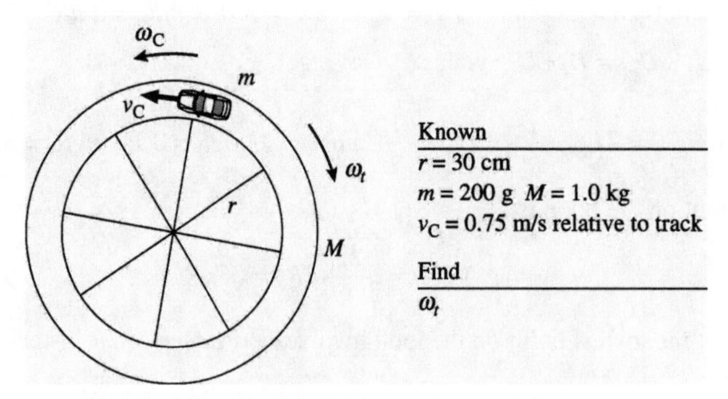


**Solve:** The toy car's steady speed of 0.75 m/s relative to the track means that

$$v_c - v_t = 0.75 \text{ m/s} \Rightarrow v_c = v_t + 0.75 \text{ m/s},$$

where $v_t$ is the velocity of a point on the track at the same radius as the car. Conservation of angular momentum implies that

$$L_i = L_f$$

$$0 = I_c\omega_c + I_t\omega_t = (mr^2)\omega_c + (Mr^2)\omega_t = m\omega_c + M\omega_t$$

The initial and final states refer to before and after the toy car was turned on. Table 12.2 was used for the track. Since $\omega_c = \frac{v_c}{r}$, $\omega_t = \frac{v_t}{r}$, we have

$$0 = mv_c + Mv_t$$

$$\Rightarrow m(v_t + 0.75 \text{ m/s}) + Mv_t = 0$$

$$\Rightarrow v_t = -\frac{M}{m+M}(0.75 \text{ m/s}) = -\frac{(0.200 \text{ kg})}{(0.200 \text{ kg} + 1.0 \text{ kg})}(0.75 \text{ m/s}) = -0.125 \text{ m/s}$$

The minus sign indicates that the track is moving in the opposite direction of the car. The angular velocity of the track is

$$\omega_t = \frac{v_t}{r} = \frac{(0.125 \text{ m/s})}{0.30 \text{ m}} = 0.417 \text{ rad/s clockwise.}$$

In rpm,

$$\omega_t = (0.417 \text{ rad/s})\left(\frac{\text{rev}}{2\pi\text{rad}}\right)\left(\frac{60 \text{ s}}{\text{min}}\right)$$

$$= 4.0 \text{ rpm}$$

**Assess:** The speed of the track is less than that of the car because it is more massive.

# 13

# NEWTON'S THEORY OF GRAVITY

**13.1. Model:** Model the sun (s) and the earth (e) as spherical masses. Due to the large difference between your size and mass and that of either the sun or the earth, a human body can be treated as a particle.

**Solve:** $F_{\text{s on you}} = \dfrac{GM_sM_y}{r_{\text{s-e}}^2}$ and $F_{\text{e on you}} = \dfrac{GM_eM_y}{r_e^2}$

Dividing these two equations gives

$$\frac{F_{\text{s on y}}}{F_{\text{e on y}}} = \left(\frac{M_s}{M_e}\right)\left(\frac{r_e}{r_{\text{s-e}}}\right)^2 = \left(\frac{1.99\times10^{30}\text{ kg}}{5.98\times10^{24}\text{ kg}}\right)\left(\frac{6.37\times10^6\text{ m}}{1.50\times10^{11}\text{ m}}\right)^2 = 6.00\times10^{-4}$$

**13.5. Model:** Model the woman (w) and the man (m) as spherical masses or particles.

**Solve:** $F_{\text{w on m}} = F_{\text{m on w}} = \dfrac{GM_wM_m}{r_{\text{m-w}}^2} = \dfrac{(6.67\times10^{-11}\text{ N}\cdot\text{m}^2/\text{kg}^2)(50\text{ kg})(70\text{ kg})}{(1.0\text{ m})^2} = 2.3\times10^{-7}\text{ N}$

**13.9. Model:** Model the earth (e) as a spherical mass.

**Visualize:** The acceleration due to gravity at sea level is $9.83\text{ m/s}^2$ (see Table 13.1) and $R_e = 6.37\times10^6\text{ m}$ (see Table 13.2).

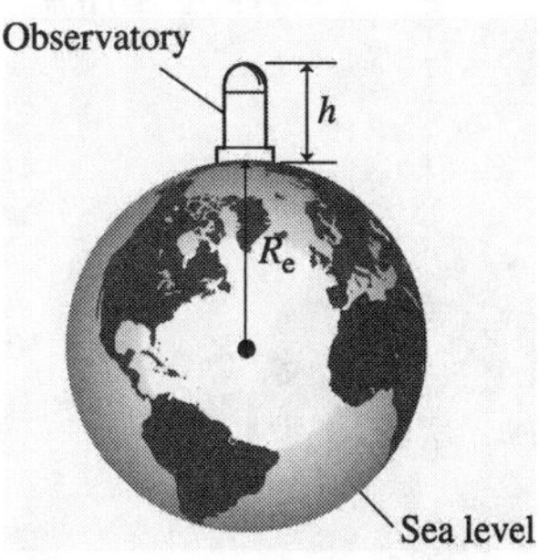


**Solve:** $$g_{\text{observatory}} = \frac{GM_e}{(R_e+h)^2} = \frac{GM_e}{R_e^2\left(1+\dfrac{h}{R_e}\right)^2} = \frac{g_{\text{earth}}}{\left(1+\dfrac{h}{R_e}\right)^2} = (9.83-0.0075)\text{ m/s}^2$$

Here $g_{\text{earth}} = GM_e/R_e^2$ is the acceleration due to gravity on a non-rotating earth, which is why we've used the value $9.83\text{ m/s}^2$. Solving for $h$,

$$h = \left(\sqrt{\frac{9.83}{9.8225}} - 1\right)R_e = 2.43\text{ km}$$

**13.21. Model:** Model the sun (S) as a spherical mass and the satellite (s) as a point particle.

**Visualize:** The satellite, having mass $m_s$ and velocity $v_s$, orbits the sun with a mass $M_S$ in a circle of radius $r_s$.

**Solve:** The gravitational force between the sun and the satellite provides the necessary centripetal acceleration for circular motion. Newton's second law is

$$\frac{GM_S m_s}{r_s^2} = \frac{m_s v_s^2}{r_s}$$

Because $v_s = 2\pi r_s / T_s$ where $T_s$ is the period of the satellite, this equation simplifies to

$$\frac{GM_S}{r_s^2} = \frac{(2\pi r_s)^2}{T_s^2 r_s} \Rightarrow r_s^3 = \frac{GM_S T_s^2}{4\pi^2} = \frac{(6.67\times10^{-11}\ \text{N}\cdot\text{m}^2/\text{kg}^2)(1.99\times10^{30}\ \text{kg})(24\times3600\ \text{s})^2}{4\pi^2} \Rightarrow r_s = 2.9\times10^9\ \text{m}$$

**13.23. Model:** Model the earth (e) as a spherical mass and the satellite (s) as a point particle.
**Visualize:** The satellite has a mass is $m_s$ and orbits the earth with a velocity $v_s$. The radius of the circular orbit is denoted by $r_s$ and the mass of the earth by $M_e$.
**Solve:** The satellite experiences a gravitational force that provides the centripetal acceleration required for circular motion:

$$\frac{GM_e m_s}{r_s^2} = \frac{m_s v_s^2}{r_s} \Rightarrow r_s = \frac{GM_e}{v_s^2} = \frac{(6.67\times10^{-11}\text{N}\cdot\text{m}^2/\text{kg}^2)(5.98\times10^{24}\ \text{kg})}{(5500\ \text{m/s})^2} = 1.32\times10^7\ \text{m}$$

$$\Rightarrow T_s = \frac{2\pi R_s}{v_s} = \frac{(2\pi)(1.32\times10^7\ \text{m})}{(5500\ \text{m/s})} = 1.51\times10^4\text{s} = 4.2\ \text{hr}$$

**13.33. Model:** Model the earth as a spherical mass and the object (o) as a point particle. Ignore air resistance. This is an isolated system, so mechanical energy is conserved.
**Visualize:**

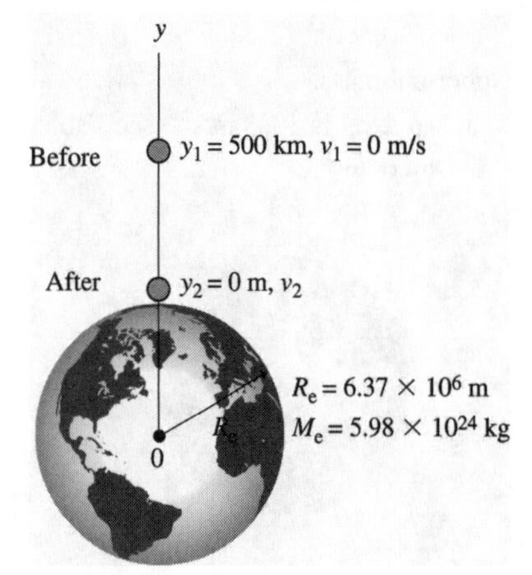


**Solve:** **(a)** The conservation of energy equation $K_2 + U_{g2} = K_1 + U_{g1}$ is

$$\frac{1}{2}m_o v_2^2 - \frac{GM_e m_o}{R_e} = \frac{1}{2}m_o v_1^2 - \frac{GM_e m_o}{(R_e + y_1)}$$

$$\Rightarrow v_2 = \sqrt{2GM_e\left(\frac{1}{R_e} - \frac{1}{R_e + y_1}\right)}$$

$$= \sqrt{2(6.67\times10^{-11}\ \text{N}\cdot\text{m}^2/\text{kg}^2)(5.98\times10^{24}\ \text{kg})\left(\frac{1}{6.37\times10^6\ \text{m}} - \frac{1}{6.87\times10^6\ \text{m}}\right)} = 3.02\ \text{km/s}$$

**(b)** In the flat-earth approximation, $U_g = mgy$. The energy conservation equation thus becomes

$$\frac{1}{2}m_o v_2^2 + m_o g y_2 = \frac{1}{2}m_o v_1^2 + m_o g y_1$$

$$\Rightarrow v_2 = \sqrt{v_1^2 + 2g(y_1 - y_2)} = \sqrt{2(9.80 \text{ m/s}^2)(5.00\times10^5 \text{ m} - 0 \text{ m})} = 3.13 \text{ km/s}$$

**(c)** The percent error in the flat-earth calculation is

$$\frac{3130 \text{ m/s} - 3020 \text{ m/s}}{3020 \text{ m/s}} \approx 3.6\%$$

**13.35. Model:** Model the planet (p) as a spherical mass and the projectile as a point mass. This is an isolated system, so mechanical energy is conserved.

**Visualize:** The projectile of mass $m$ was launched on the surface of the planet with an initial velocity $v_0$.

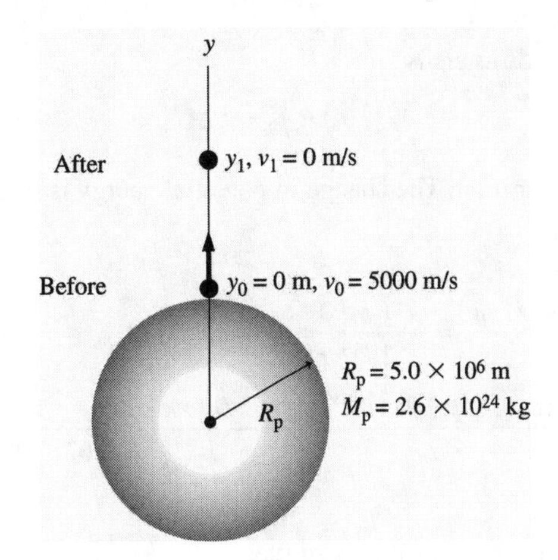


**Solve:** **(a)** The energy conservation equation $K_1 + U_1 = K_0 + U_0$ is

$$\frac{1}{2}mv_1^2 - \frac{GM_p m}{R_p + y_1} = \frac{1}{2}mv_0^2 - \frac{GM_p m}{R_p}$$

$$\Rightarrow y_1 = \left[\frac{1}{R_p} - \frac{v_0^2}{2GM_p}\right]^{-1} - R_p = 2.8\times10^6 \text{ m}$$

**(b)** Using the energy conservation equation $K_1 + U_1 = K_0 + U_0$ with $y_1 = 1000 \text{ km} = 1.000\times10^6$ m:

$$\frac{1}{2}mv_1^2 - \frac{GM_p m}{R_p + y_1} = \frac{1}{2}mv_0^2 - \frac{GM_p m}{R_p}$$

$$\Rightarrow v_1 = \left[v_0^2 + 2GM_p\left(\frac{1}{R_p + y_1} - \frac{1}{R_p}\right)\right]^{1/2}$$

$$= \left[(5000 \text{ m/s})^2 + 2(6.67\times10^{-11} \text{N}\cdot\text{m}^2/\text{kg}^2)(2.6\times10^{24} \text{ kg})\left(\frac{1}{(5.0\times10^6 \text{ m} + 1.000\times10^6 \text{ m})} - \frac{1}{5.0\times10^6 \text{ m}}\right)\right]^{1/2}$$

$$= 3.7 \text{ km/s}$$

**13.47. Model:** Model the earth (e) as a spherical mass and the space shuttle (s) as a point particle. This is an isolated system, so the mechanical energy is conserved.

**Visualize:** The space shuttle $(\text{mass} = m_s)$ is at a distance of $r_1 = R_e + 250$ km.

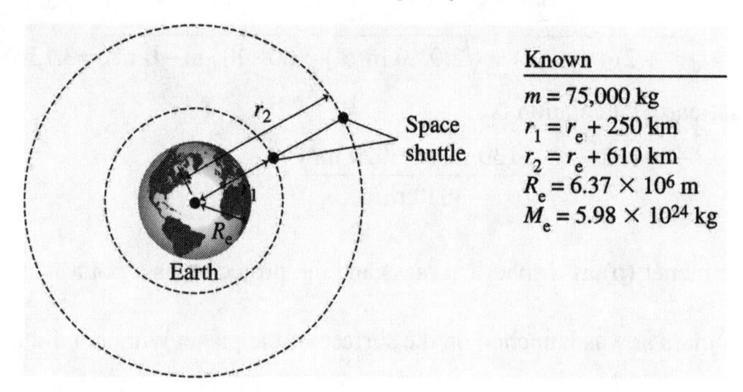


**Solve:** The external work done by the thrusters is

$$W_{ext} = \Delta E_{mech} = \tfrac{1}{2}\Delta U_g$$

where we used $E_{mech} = \tfrac{1}{2}U_g$ for a circular orbit. The change in potential energy is from the initial orbit at $r_1 = R_e + 250$ km to the final orbit $r_f = R_e + 610$ km. Thus

$$W_{ext} = \frac{1}{2}\left(\frac{-GM_em}{r_f} - \frac{-GM_em}{r_i}\right) = \frac{GM_em}{2}\left(\frac{1}{r_i} - \frac{1}{r_f}\right)$$

$$= \frac{(6.67\times10^{-11}\ \text{N m}^2/\text{kg}^2)(5.98\times10^{24}\ \text{kg})(75{,}000\ \text{kg})}{2}\left(\frac{1}{6.62\times10^6\ \text{m}} - \frac{1}{6.98\times10^6\ \text{m}}\right)$$

$$= 1.17\times10^{11}\ \text{J}$$

This much energy must be supplied by burning the on-board fuel.

**13.49. Model:** Planet Physics is a spherical mass. The cruise ship is in a circular orbit.

**Solve:** **(a)** At the surface, the free-fall acceleration is $g = G\dfrac{M}{R^2}$. From kinematics,

$$y_f = y_i + v_0\Delta t - 2g(\Delta t)^2 \Rightarrow 0\ \text{m} = 0\ \text{m} + (11\ \text{m/s})(2.5\ \text{s}) - 2g(2.5\ \text{s})^2$$

$$\Rightarrow g = 2.20\ \text{m/s}^2$$

The period of the cruise ship's orbit is $230\times60 = 13{,}800$ s. For the circular orbit of the cruise ship,

$$T^2 = \left(\frac{4\pi^2}{GM}\right)(2R)^3 \Rightarrow \frac{T^2}{32\pi^2} = R\left(\frac{R^2}{GM}\right) = R\left(\frac{1}{g}\right)$$

$$\Rightarrow R = \frac{(2.20\ \text{m/s}^2)(13{,}800\ \text{s})^2}{32\pi^2} = 1.327\times10^6\ \text{m}.$$

The mass is thus $M = \dfrac{R^2}{G}g = 5.8\times10^{22}$ kg.

**(b)** From part (a), $R = 1.33\times10^6$ m.

**13.53. Solve:** **(a)** Taking the logarithm of both sides of $v^p = Cu^q$ gives

$$[\log(v^p) = p\log v] = [\log(Cu^q) = \log C + q\log u] \Rightarrow \log v = \frac{q}{p}\log u + \frac{\log C}{p}$$

But $x = \log u$ and $y = \log v$, so $x$ and $y$ are related by

$$y = \left(\frac{q}{p}\right)x + \frac{\log C}{p}$$

**(b)** The previous result shows there is a linear relationship between $x$ and $y$, hence there is a linear relationship between log $u$ and log $v$. The graph of a linear relationship is a straight line, so the graph of log $v$-versus-log $u$ will be a straight line.

**(c)** The slope of the straight line represented by the equation $y = (q/p)x + \log C/p$ is $q/p$. Thus, the slope of the log $v$-versus-log $u$ graph will be $q/p$.

**(d)** From Newton's theory, the period $T$ and radius $r$ of an orbit around the sun are related by

$$T^2 = \left(\frac{4\pi^2}{GM}\right)r^3$$

This equation is of the form $T^p = Cr^q$, with $p = 2$, $q = 3$, and $C = 4\pi^2/GM$. If the theory is correct, we *expect* a graph of log $T$-versus-log $r$ to be a straight line with slope $q/p = 3/2 = 1.500$. The *experimental measurements* of actual planets yield a straight line graph whose slope is 1.500 to four significant figures. Note that the graph has nothing to do with theory—it is simply a graph of measured values. But the fact that the shape and slope of the graph agree precisely with the prediction of Newton's theory is strong evidence for its correctness.

**(e)** The predicted $y$-intercept of the graph is log $C/p$, and the experimentally determined value is 9.264. Equating these, we can solve for $M$. Because the planets all orbit the sun, the mass we are finding is $M = M_{\text{sun}}$.

$$\frac{1}{2}\log C = \frac{1}{2}\log\left(\frac{4\pi^2}{GM_{\text{sun}}}\right) = -9.264 \Rightarrow \frac{4\pi^2}{GM_{\text{sun}}} = 10^{-18.528} = \frac{1}{10^{18.528}}$$

$$\Rightarrow M_{\text{sun}} = \frac{4\pi^2}{G}\cdot 10^{18.528} = 1.996\times 10^{30}\ \text{kg}$$

The tabulated value, to three significant figures, is $M_{\text{sum}} = 1.99\times 10^{30}$ kg. We have used the orbits of the planets to "weigh the sun!"

**13.59. Model:** Angular momentum is conserved for a particle following a trajectory of any shape.
**Visualize:**

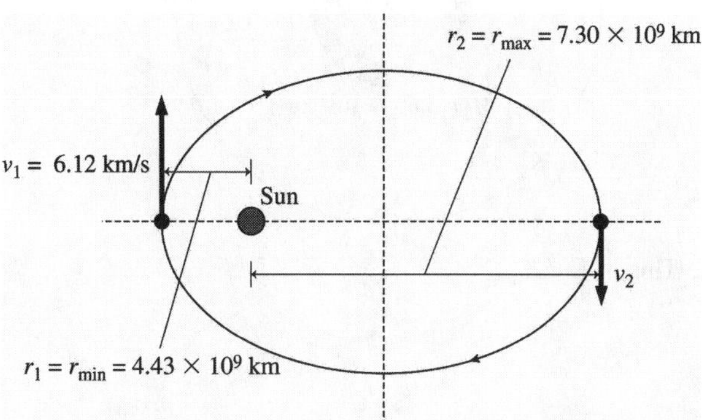


For a particle in an elliptical orbit, the particle's angular momentum is $L = mrv_t = mrv\sin\beta$, where $v$ is the velocity tangent to the trajectory and $\beta$ is the angle between $\vec{r}$ and $\vec{v}$.

**Solve:** At the distance of closest approach ($r_{\text{min}}$) and also at the most distant point, $\beta = 90°$. Since there is no tangential force (the only force being the radial force), angular momentum must be conserved:

$$m_{\text{Pluto}} v_1 r_{\text{min}} = m_{\text{Pluto}} v_2 r_{\text{max}}$$

$$\Rightarrow v_2 = v_1(r_{\text{min}}/r_{\text{max}}) = (6.12\times 10^3\ \text{m/s})(4.43\times 10^{12}\ \text{m}/7.30\times 10^{12}\ \text{m}) = 3.71\ \text{km/s}$$

**13.61. Model:** For the sun + comet system, the mechanical energy is conserved.

**Visualize:**

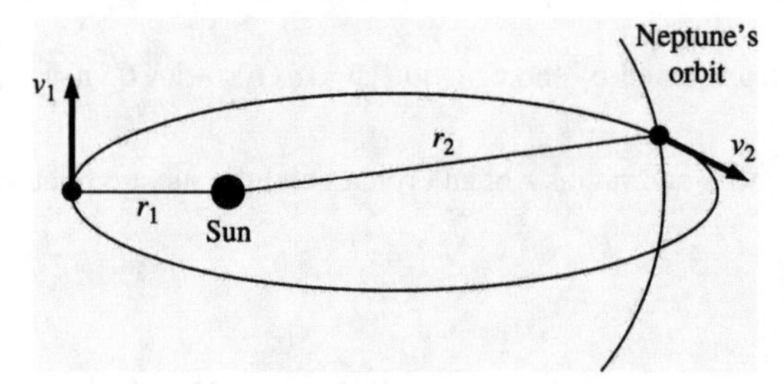


**Solve:** The conservation of energy equation $K_f + U_f = K_i + U_i$ is

$$\frac{1}{2}M_c v_2^2 - \frac{GM_s M_c}{r_2} = \frac{1}{2}M_c v_1^2 - \frac{GM_s M_c}{r_1}$$

Using $G = 6.67\times10^{-11}\ \text{Nm}^2/\text{kg}^2$, $M_s = 1.99\times10^{30}$ kg, $r_1 = 8.79\times10^{10}$ m, $r_2 = 4.50\times10^{12}$ m, and $v_1 = 54.6$ km/s, we get $v_2 = 4.49$ km/s.

**13.65. Solve:** **(a)** At what distance from the center of Saturn is the acceleration due to gravity the same as on the surface of the earth?

**(b)**

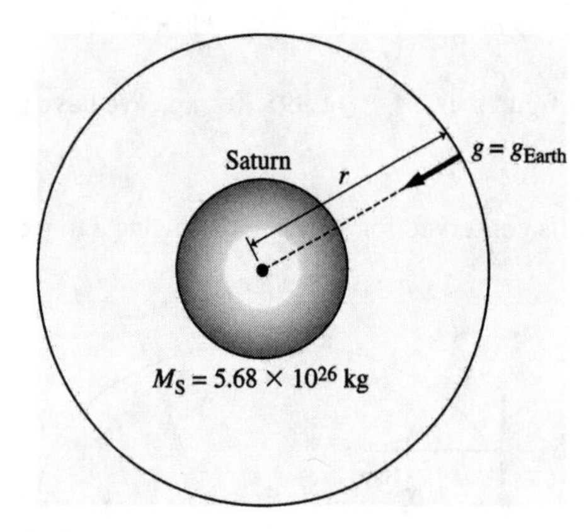


(c) The distance is $6.21\times10^7$ m. This is $1.06R_{\text{Saturn}}$

# OSCILLATIONS

# 14

**14.3. Model:** The air-track glider attached to a spring is in simple harmonic motion.
**Visualize:** The position of the glider can be represented as $x(t) = A\cos\omega t$.

**Solve:** The glider is pulled to the right and released from rest at $t = 0$ s. It then oscillates with a period $T = 2.0$ s and a maximum speed $v_{max} = 40\text{ cm/s} = 0.40\text{ m/s}$.

**(a)** $v_{max} = \omega A$ and $\omega = \dfrac{2\pi}{T} = \dfrac{2\pi}{2.0\text{ s}} = \pi\text{ rad/s} \Rightarrow A = \dfrac{v_{max}}{\omega} = \dfrac{0.40\text{ m/s}}{\pi\text{ rad/s}} = 0.127\text{ m} = 12.7\text{ cm}$

**(b)** The glider's position at $t = 0.25$ s is

$$x_{0.25\text{ s}} = (0.127\text{ m})\cos\left[(\pi\text{ rad/s})(0.25\text{ s})\right] = 0.090\text{ m} = 9.0\text{ cm}$$

**14.9. Solve:** The position of the object is given by the equation

$$x(t) = A\cos(\omega t + \phi_0)$$

The amplitude $A = 8.0$ cm. The angular frequency $\omega = 2\pi f = 2\pi(0.50\text{ Hz}) = \pi$ rad/s. Since at $t = 0$ it has its most negative velocity, it must be at the equilibrium point $x = 0$ cm and moving to the left, so $\phi_0 = \dfrac{\pi}{2}$. Thus

$$x(t) = (8.0\text{ cm})\cos[(\pi\text{ rad/s})t + \frac{\pi}{2}\text{ rad}]$$

**14.13. Model:** The mass attached to the spring oscillates in simple harmonic motion.
**Solve:** **(a)** The period $T = 1/f = 1/2.0\text{ Hz} = 0.50\text{ s}$.
**(b)** The angular frequency $\omega = 2\pi f = 2\pi(2.0\text{ Hz}) = 4\pi\text{ rad/s}$.
**(c)** Using energy conservation

$$\tfrac{1}{2}kA^2 = \tfrac{1}{2}kx_0^2 + \tfrac{1}{2}mv_{0x}^2$$

Using $x_0 = 5.0$ cm, $v_{0x} = -30$ cm/s and $k = m\omega^2 = (0.200\text{ kg})(4\pi\text{ rad/s})^2$, we get $A = 5.54$ cm.
**(d)** To calculate the phase constant $\phi_0$,

$$A\cos\phi_0 = x_0 = 5.0\text{ cm}$$

$$\Rightarrow \phi_0 = \cos^{-1}\left(\frac{5.0\text{ cm}}{5.54\text{ cm}}\right) = 0.45\text{ rad}$$

**(e)** The maximum speed is $v_{max} = \omega A = (4\pi\text{ rad/s})(5.54\text{ cm}) = 70\text{ cm/s}$.
**(f)** The maximum acceleration is

$$a_{max} = \omega^2 A = \omega(\omega A) = (4\pi\text{ rad/s})(70\text{ cm/s}) = 8.8\text{ m/s}^2$$

**(g)** The total energy is $E = \tfrac{1}{2}mv_{max}^2 = \tfrac{1}{2}(0.200\text{ kg})(0.70\text{ m/s})^2 = 0.049$ J.

**(h)** The position at $t = 0.40$ s is

$$x_{0.4\text{ s}} = (5.54\text{ cm})\cos\left[(4\pi\text{ rad/s})(0.40\text{ s}) + 0.45\text{ rad}\right] = +3.8\text{ cm}$$

**14.15. Model:** The block attached to the spring is in simple harmonic motion.
**Visualize:**

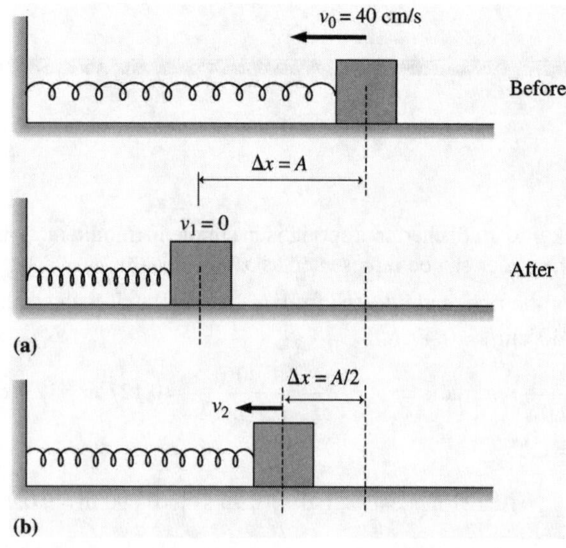


**Solve:** **(a)** The conservation of mechanical energy equation $K_f + U_{sf} = K_i + U_{si}$ is

$$\tfrac{1}{2}mv_1^2 + \tfrac{1}{2}k(\Delta x)^2 = \tfrac{1}{2}mv_0^2 + 0\text{ J} \Rightarrow 0\text{ J} + \tfrac{1}{2}kA^2 = \tfrac{1}{2}mv_0^2 + 0\text{ J}$$

$$\Rightarrow A = \sqrt{\frac{m}{k}}v_0 = \sqrt{\frac{1.0\text{ kg}}{16\text{ N/m}}}(0.40\text{ m/s}) = 0.10\text{ m} = 10.0\text{ cm}$$

**(b)** We have to find the velocity at a point where $x = A/2$. The conservation of mechanical energy equation $K_2 + U_{s2} = K_i + U_{si}$ is

$$\frac{1}{2}mv_2^2 + \frac{1}{2}k\left(\frac{A}{2}\right)^2 = \frac{1}{2}mv_0^2 + 0\text{ J} \Rightarrow \frac{1}{2}mv_2^2 = \frac{1}{2}mv_0^2 - \frac{1}{4}\left(\frac{1}{2}kA^2\right) = \frac{1}{2}mv_0^2 - \frac{1}{4}\left(\frac{1}{2}mv_0^2\right) = \frac{3}{4}\left(\frac{1}{2}mv_0^2\right)$$

$$\Rightarrow v_2 = \sqrt{\frac{3}{4}}v_0 = \sqrt{\frac{3}{4}}(0.40\text{ m/s}) = 0.346\text{ m/s}$$

The velocity is 35 cm/s.

**14.17. Model:** The vertical oscillations constitute simple harmonic motion.
**Solve:** To find the oscillation frequency using $\omega = 2\pi f = \sqrt{k/m}$, we first need to find the spring constant $k$. In equilibrium, the weight $mg$ of the block and the spring force $k\Delta L$ are equal and opposite. That is, $mg = k\Delta L \Rightarrow k = mg/\Delta L$. The frequency of oscillation $f$ is thus given as

$$f = \frac{1}{2\pi}\sqrt{\frac{k}{m}} = \frac{1}{2\pi}\sqrt{\frac{mg/\Delta L}{m}} = \frac{1}{2\pi}\sqrt{\frac{g}{\Delta L}} = \frac{1}{2\pi}\sqrt{\frac{9.8\text{ m/s}^2}{0.020\text{ m}}} = 3.5\text{ Hz}$$

**14.27. Model:** The motion is a damped oscillation.
**Solve:** The amplitude of the oscillation at time $t$ is given by Equation 14.58: $A(t) = A_0e^{-t/2\tau}$, where $\tau = m/b$ is the time constant. Using $x = 0.368$ A and $t = 10.0$ s, we get

$$0.368A = Ae^{-10.0\text{ s}/2\tau} \Rightarrow \ln(0.368) = \frac{-10\text{ s}}{2\tau} \Rightarrow \tau = -\frac{10.0\text{ s}}{2\ln(0.368)} = 5.00\text{ s}$$

**14.35. Model:** The block attached to the spring is in simple harmonic motion.
**Visualize:** The position and the velocity of the block are given by the equations

$$x(t) = A\cos(\omega t + \phi_0) \text{ and } v_x(t) = -A\omega\sin(\omega t + \phi_0)$$

**Solve:** To graph *x*(*t*) we need to determine $\omega$, $\phi_0$, and *A*. These quantities will be found by using the initial ($t = 0$ s) conditions on *x*(*t*) and $v_x(t)$. The period is

$$T = 2\pi\sqrt{\frac{m}{k}} = 2\pi\sqrt{\frac{1.0 \text{ kg}}{20 \text{ N/m}}} = 1.405 \text{ s} \Rightarrow \omega = \frac{2\pi}{T} = \frac{2\pi \text{ rad}}{1.405 \text{ s}} = 4.472 \text{ rad/s}$$

At $t = 0$ s, $x_0 = A\cos\phi_0$ and $v_{0x} = -A\omega\sin\phi_0$. Dividing these equations,

$$\tan\phi_0 = -\frac{v_{0x}}{\omega x_0} = -\frac{(-1.0 \text{ m/s})}{(4.472 \text{ rad/s})(0.20 \text{ m})} = 1.1181 \Rightarrow \phi_0 = 0.841 \text{ rad}$$

From the initial conditions,

$$A = \sqrt{x_0^2 + \left(\frac{v_{0x}}{\omega}\right)^2} = \sqrt{(0.20 \text{ m})^2 + \left(\frac{-1.0 \text{ m/s}}{4.472 \text{ rad/s}}\right)^2} = 0.300 \text{ m}$$

The position-versus-time graph can now be plotted using the equation

$$x(t) = (0.300 \text{ m})\cos[(4.472 \text{ rad/s})t + 0.841 \text{ rad}]$$

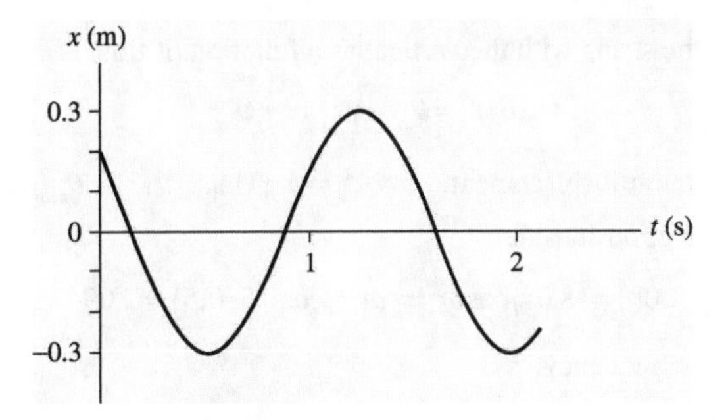


**14.37. Model:** The particle is in simple harmonic motion.
**Solve:** The equation for the velocity of the particle is

$$v_x(t) = -(25 \text{ cm})(10 \text{ rad/s})\sin(10 \text{ t})$$

Substituting into $K = 2U$ gives

$$\frac{1}{2}mv_x^2(t) = 2\left(\frac{1}{2}kx^2(t)\right) \Rightarrow \frac{1}{2}m[-(250 \text{ cm/s})\sin(10 \text{ t})]^2 = k[(25 \text{ cm})\cos(10 \text{ t})]^2$$

$$\Rightarrow \frac{\sin^2(10 \text{ t})}{\cos^2(10 \text{ t})} = 2\left(\frac{k}{m}\right)\frac{(25 \text{ cm})^2}{(250 \text{ cm/s})^2} = 2\omega^2\left(\frac{1}{100}\right) \text{s}^2$$

$$\Rightarrow \tan^2(10 \text{ t}) = 2(10 \text{ rad/s})^2\left(\frac{1}{100}\right) \text{s}^2 = 2.0 \Rightarrow t = \frac{1}{10}\tan^{-1}\sqrt{2.0} = 0.096 \text{ s}$$

**14.41. Model:** The block on a spring is in simple harmonic motion.

**Solve:** **(a)** The position of the block is given by $x(t) = A\cos(\omega t + \phi_0)$. Because $x(t) = A$ at $t = 0$ s, we have $\phi_0 = 0$ rad, and the position equation becomes $x(t) = A\cos\omega t$. At $t = 0.685$ s, $3.00 \text{ cm} = A\cos(0.685\omega)$ and at $t = 0.886$ s, $-3.00 \text{ cm} = A\cos(0.886\omega)$. These two equations give

$$\cos(0.685\omega) = -\cos(0.886\omega) = \cos(\pi - 0.886\omega)$$
$$\Rightarrow 0.685\omega = \pi - 0.886\omega \Rightarrow \omega = 2.00 \text{ rad/s}$$

**(b)** Substituting into the position equation,

$$3.00\text{ cm} = A\cos\left((2.00\text{ rad/s})(0.685\text{ s})\right) = A\cos(1.370) = 0.20A \Rightarrow A = \frac{3.00\text{ cm}}{0.20} = 15.0\text{ cm}$$

**14.49. Model:** The compact car is in simple harmonic motion.

**Solve:** **(a)** The mass on each spring is $(1200\text{ kg})/4 = 300\text{ kg}$. The spring constant can be calculated as follows:

$$\omega^2 = \frac{k}{m} \Rightarrow k = m\omega^2 = m(2\pi f)^2 = (300\text{ kg})\left[2\pi(2.0\text{ Hz})\right]^2 = 4.74\times10^4\text{ N/m}$$

The spring constant is $4.7\times10^4$ N/m.

**(b)** The car carrying four persons means that each spring has, on the average, an additional mass of 70 kg. That is, $m = 300\text{ kg} + 70\text{ kg} = 370\text{ kg}$. Thus,

$$f = \frac{\omega}{2\pi} = \frac{1}{2\pi}\sqrt{\frac{k}{m}} = \frac{1}{2\pi}\sqrt{\frac{4.74\times10^4\text{ N/m}}{370\text{ kg}}} = 1.80\text{ Hz}$$

**Assess:** A small frequency change from the additional mass is reasonable because frequency is inversely proportional to the square root of the mass.

**14.55. Model:** Assume that the angle with the vertical that the pendulum makes is small enough so that there is simple harmonic motion.

**Solve:** The angle $\theta$ made by the string with the vertical as a function of time is

$$\theta(t) = \theta_{\text{max}}\cos(\omega t + \phi_0)$$

The pendulum starts from maximum displacement, thus $\phi_0 = 0$. Thus, $\theta(t) = \theta_{\text{max}}\cos\omega t$. To find the time $t$ when the pendulum reaches 4.0° on the opposite side:

$$(-4.0°) = (8.0°)\cos\omega t \Rightarrow \omega t = \cos^{-1}(-0.5) = 2.094\text{ rad}$$

Using the formula for the angular frequency,

$$\omega = \sqrt{\frac{g}{L}} = \sqrt{\frac{9.8\text{ m/s}^2}{1.0\text{ m}}} = 3.130\text{ rad/s} \Rightarrow t = \frac{2.0944\text{ rad}}{\omega} = \frac{2.094\text{ rad}}{3.130\text{ rad/s}} = 0.669\text{ s}$$

The time $t = 0.67$ s.

**Assess:** Because $T = 2\pi/\omega = 2.0$ s, a value of 0.67 s for the pendulum to cover a little less than half the oscillation is reasonable.

**14.61. Model:** A completely inelastic collision between the two gliders resulting in simple harmonic motion.
**Visualize:**

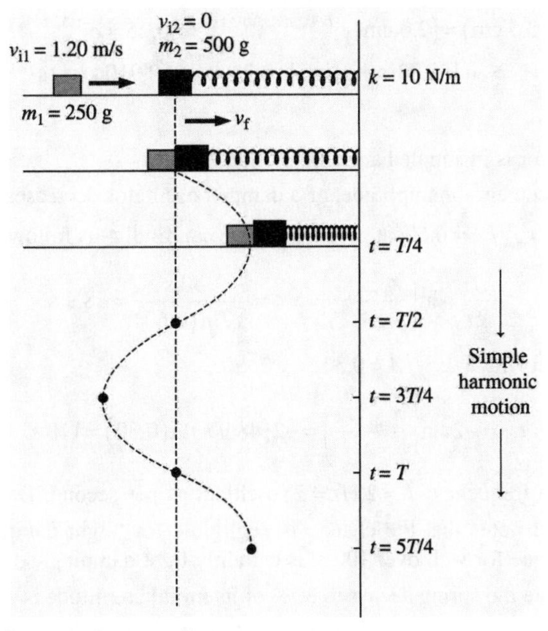


Let us denote the 250 g and 500 g masses as $m_1$ and $m_2$, which have initial velocities $v_{i1}$ and $v_{i2}$. After $m_1$ collides with and sticks to $m_2$, the two masses move together with velocity $v_f$.

**Solve:** The momentum conservation equation $p_f = p_i$ for the completely inelastic collision is $(m_1 + m_2)v_f = m_1 v_{i1} + m_2 v_{i2}$. Substituting the given values,

$$(0.750\text{ kg})v_f = (0.250\text{ kg})(1.20\text{ m/s}) + (0.500\text{ kg})(0\text{ m/s}) \Rightarrow v_f = 0.400\text{ m/s}$$

We now use the conservation of mechanical energy equation:

$$(K + U_s)_{\text{compressed}} = (K + U_s)_{\text{equilibrium}} \Rightarrow 0\text{ J} + \tfrac{1}{2}kA^2 = \tfrac{1}{2}(m_1 + m_2)v_f^2 + 0\text{ J}$$

$$\Rightarrow A = \sqrt{\frac{m_1 + m_2}{k}}v_f = \sqrt{\frac{0.750\text{ kg}}{10\text{ N/m}}}(0.400\text{ m/s}) = 0.110\text{ m}$$

The period is

$$T = 2\pi\sqrt{\frac{m_1 + m_2}{k}} = 2\pi\sqrt{\frac{0.750\text{ kg}}{10\text{ N/m}}} = 1.72\text{ s}$$

**14.69. Model:** The doll's head is in simple harmonic motion and is damped.
**Solve:** **(a)** The oscillation frequency is

$$f = \frac{1}{2\pi}\sqrt{\frac{k}{m}} \Rightarrow k = m(2\pi f)^2 = (0.015\text{ kg})(2\pi)^2(4.0\text{ Hz})^2 = 9.475\text{ N/m}$$

The spring constant is 9.5 N/m.
**(b)** The maximum speed is

$$v_{\max} = \omega A = \sqrt{\frac{k}{m}}A = \sqrt{\frac{9.475\text{ N/m}}{0.015\text{ kg}}}(0.020\text{ m}) = 0.50\text{ m/s}$$

**(c)** Using $A(t) = A_0 e^{-bt/2m}$, we get

$$(0.5\text{ cm}) = (2.0\text{ cm})e^{-b(4.0\text{ s})/(2\times 0.015\text{ kg})} \Rightarrow 0.25 = e^{-(133.3\text{ s/kg})b}$$
$$\Rightarrow -(133.33\text{ s/kg})b = \ln 0.25 \Rightarrow b = 0.0104\text{ kg/s}$$

**14.71. Model:** The oscillator is in simple harmonic motion.

**Solve:** The maximum displacement, or amplitude, of a damped oscillator decreases as $x_{max}(t) = Ae^{-t/2\tau}$, where $\tau$ is the time constant. We know $x_{max}/A = 0.60$ at $t = 50$ s, so we can find $\tau$ as follows:

$$-\frac{t}{2\tau} = \ln\left(\frac{x_{max}(t)}{A}\right) \Rightarrow \tau = -\frac{50\text{ s}}{2\ln(0.60)} = 48.9\text{ s}$$

Now we can find the time $t_{30}$ at which $x_{max}/A = 0.30$:

$$t_{30} = -2\tau\ln\left(\frac{x_{max}(t)}{A}\right) = -2(48.9\text{ s})\ln(0.30) = 118\text{ s}$$

The undamped oscillator has a frequency $f = 2\text{ Hz} = 2$ oscillations per second. Damping changes the oscillation frequency slightly, but the text notes that the change is negligible for "light damping." Damping by air, which allows the oscillations to continue for well over 100 s, is certainly light damping, so we will use $f = 2.0$ Hz. Then the number of oscillations before the spring decays to 30% of its initial amplitude is

$$N = f \cdot t_{30} = (2\text{ oscillations/s}) \cdot (118\text{ s}) = 236\text{ oscillations}$$

# 15

## FLUIDS AND ELASTICITY

**15.3. Model:** The density of water is 1000 kg/m$^3$.

**Visualize:**

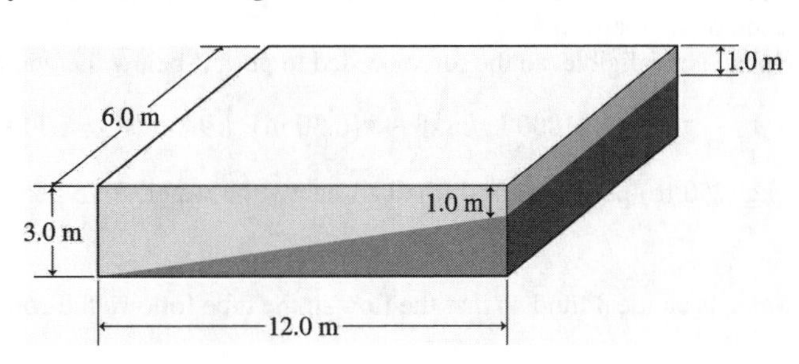


**Solve:** Volume of water in the swimming pool is

$$V = 6\text{ m}\times 12\text{ m}\times 3\text{ m} - \tfrac{1}{2}(6\text{ m}\times 12\text{ m}\times 2\text{ m}) = 144\text{ m}^3$$

The mass of water in the swimming pool is

$$m = \rho V = (1000\text{ kg/m}^3)(144\text{ m}^3) = 1.44\times 10^5\text{ kg}$$

**15.5. Model:** The density of sea water is 1030 kg/m$^3$.

**Solve:** The pressure below sea level can be found from Equation 15.6 as follows:

$$p = p_0 + \rho g d = 1.013\times 10^5\text{Pa} + (1030\text{ kg/m}^3)(9.80\text{ m/s}^2)(1.1\times 10^4\text{ m})$$
$$= 1.013\times 10^5\text{ Pa} + 1.1103\times 10^8\text{ Pa} = 1.1113\times 10^8\text{ Pa} = 1.10\times 10^3\text{ atm}$$

where we have used the conversion 1 atm $= 1.013\times 10^5$ Pa.

**Assess:** The pressure deep in the ocean is very large.

**15.9. Model:** The density of seawater $\rho_{\text{seawater}} = 1030\text{ kg/m}^2$.

**Visualize:**

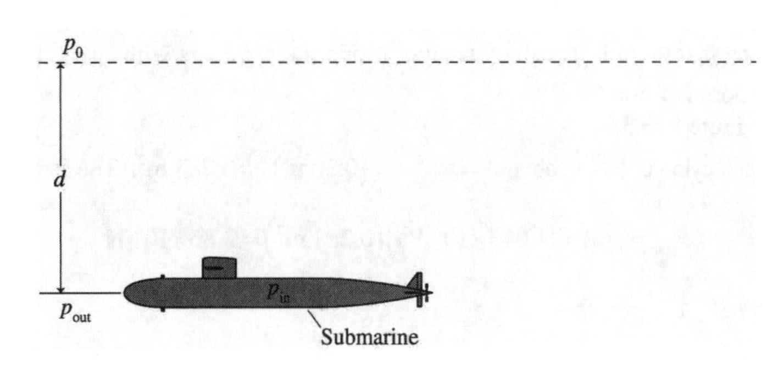

**Solve:** The pressure outside the submarine's window is $p_{out} = p_0 + \rho_{seawater} gd$, where *d* is the maximum safe depth for the window to withstand a force *F*. This force is $F/A = p_{out} - p_{in}$, where *A* is the area of the window. With $p_{in} = p_0$, we simplify the pressure equation to

$$p_{out} - p_0 = \frac{F}{A} = \rho_{seawater} gd \Rightarrow d = \frac{F}{A\rho_{seawater} g} \qquad d = \frac{1.0\times10^6 \text{ N}}{\pi(0.10 \text{ m})^2(1030 \text{ kg/m}^2)(9.8 \text{ m/s}^2)} = 3.2 \text{ km}$$

**Assess:** A force of $1.0\times10^6$ N corresponds to a pressure of

$$p = \frac{F}{A} = \frac{1.0\times10^6 \text{ N}}{\pi(0.10 \text{ m})^2} = 314 \text{ atm}$$

A depth of 3 km is therefore reasonable.

**15.21. Model:** The buoyant force is determined by Archimedes' principle. Ignore any compression the air in the beach ball may undergo as a result of submersion.
**Solve:** The mass of the beach ball is negligible, so the force needed to push it below the water is equal to the buoyant force.

$$F_B = p_w\left(\frac{4}{3}\pi R^3\right)g = (1000 \text{ kg/m}^3)\left(\frac{4}{3}\pi(0.30 \text{ m})^3\right)(9.8 \text{ m/s}^2) = 1.11 \text{ kN}$$

**Assess:** It would take a 113 kg (250 lb) person to push the ball below the water. Two people together could do it. This seems about right.

**15.25. Model:** Treat the water as an ideal fluid so that the flow in the tube follows the continuity equation.
**Visualize:**

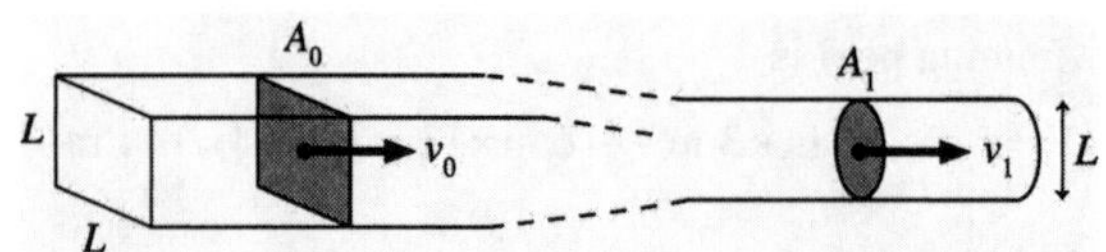


**Solve:** The equation of continuity is $v_0 A_0 = v_1 A_1$, where $A_0 = L^2$ and $A_1 = \pi(\frac{1}{2}L)^2$. The above equation simplifies to

$$v_0 L^2 = v_1 \pi\left(\frac{L}{2}\right)^2 \Rightarrow v_1 = \left(\frac{4}{\pi}\right)v_0 = 1.27 v_0$$

**15.29. Model:** The hanging mass creates tensile stress in the wire.
**Solve:** The force (*F*) pulling on the wire, which is simply the gravitational force (*mg*) on the hanging mass, produces tensile stress given by *F*/*A*, where *A* is the cross-sectional area of the wire. From the definition of Young's modulus, we have

$$Y = \frac{mg/A}{\Delta L/L} \Rightarrow m = \frac{(\pi r^2)Y\Delta L}{gL} = \frac{\pi(2.50\times10^{-4} \text{ m})^2(20\times10^{10} \text{ N/m}^2)(1.0\times10^{-3} \text{ m})}{(9.80 \text{ m/s}^2)(2.0 \text{ m})} = 2.0 \text{ kg}$$

**15.33. Model:** We assume that there is a perfect vacuum inside the cylinders with $p = 0$ Pa. We also assume that the atmospheric pressure in the room is 1 atm.
**Visualize:** Please refer to Figure P15.33.
**Solve:** (a) The flat end of each cylinder has an area $A = \pi r^2 = \pi(0.30 \text{ m})^2 = 0.283 \text{ m}^2$. The force on each end is thus

$$F_{atm} = p_0 A = (1.013\times10^5 \text{ Pa})(0.283 \text{ m}^2) = 2.86\times10^4 \text{ N}$$

The force on each end is $2.9\times10^4$ N.

**(b)** The net vertical force on the lower cylinder when it is on the verge of being pulled apart is

$$\sum F_y = F_{\text{atm}} - (F_G)_{\text{players}} = 0\ \text{N} \Rightarrow (F_G)_{\text{players}} = F_{\text{atm}} = 2.86\times10^4\ \text{N}$$

$$\Rightarrow \text{number of players} = \frac{2.86\times10^4\ \text{N}}{(100\ \text{kg})(9.8\ \text{m/s}^2)} = 29.2$$

That is, 30 players are needed to pull the two cylinders apart.

**15.37. Solve:** The fact that atmospheric pressure at sea level is $101.3\ \text{kPa} = 101{,}300\ \text{N/m}^2$ means that the weight of the atmosphere over each square meter of surface is 101,300 N. Thus the mass of air over each square meter is $m =$ $(101{,}300\ \text{N})/g = (101{,}300\ \text{N})/(9.80\ \text{m/s}^2) = 10{,}340\ \text{kg per m}^2$. Multiplying by the earth's surface area will give the total mass. Using $R_e = 6.27\times10^6\ \text{m}$ for the earth's radius, the total mass of the atmosphere is

$$M_{\text{air}} = A_{\text{earth}}m = (4\pi R_e^2)m = 4\pi(6.37\times10^6\ \text{m})^2(10{,}340\ \text{kg/m}^2) = 5.27\times10^{18}\ \text{kg}$$

**15.41. Model:** Assume that oil is incompressible and its density is $900\ \text{kg/m}^3$.
**Visualize:** Please refer to Figure P15.41.
**Solve:** **(a)** The hydraulic lift is in equilibrium and the pistons on the left and the right are at the same level. Equation 15.11, therefore, simplifies to

$$\frac{F_{\text{left piston}}}{A_{\text{left piston}}} = \frac{F_{\text{right piston}}}{A_{\text{right piston}}} \Rightarrow \frac{(F_G)_{\text{student}}}{\pi(r_{\text{student}})^2} = \frac{(F_G)_{\text{elephant}}}{\pi(r_{\text{elephant}})^2}$$

$$\Rightarrow r_{\text{student}} = \sqrt{\left(\frac{(F_G)_{\text{student}}}{(F_G)_{\text{elephant}}}\right)}(r_{\text{elephant}}) = \sqrt{\frac{(70\ \text{kg})g}{(1200\ \text{kg})g}}(1.0\ \text{m}) = 0.2415\ \text{m}$$

The diameter of the piston the student is standing on is therefore $2\times0.2415\ \text{m} = 0.48\ \text{m}$.
**(b)** From Equation 15.13, we see that an additional force $\Delta F$ is required to increase the elephant's elevation through a distance $d_2$. That is,

$$\Delta F = \rho g\left(A_{\text{left piston}} + A_{\text{right piston}}\right)d_2$$

$$\Rightarrow (70\ \text{kg})(9.8\ \text{m/s}^2) = (900\ \text{kg/m}^3)(9.8\ \text{m/s}^2)\pi\left[(0.2415\ \text{m})^2 + (1.0\ \text{m})^2\right]d_2$$

$$\Rightarrow d_2 = 0.0234\ \text{m}$$

The elephant moves 2.3 cm.

**15.47. Visualize:**

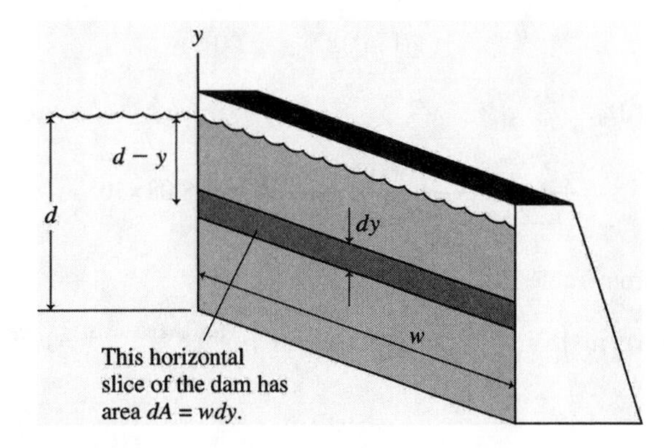


The figure shows a dam with water height *d*. We chose a coordinate system with the origin at the bottom of the dam. The horizontal slice has height *dy*, width *w*, and area $dA = wdy$. The slice is at the depth of $d - y$.

**Solve:** **(a)** The water exerts a small force $dF = pdA = pwdy$ on this small piece of the dam, where $p = \rho g(d-y)$ is the pressure at depth $d-y$. Altogether, the force on this small horizontal slice at position $y$ is $dF = \rho gw(d-y)dy$. Note that a force from atmospheric pressure is not included. This is because atmospheric pressure exerts a force on *both* sides of the dam. The total force *of the water* on the dam is found by adding up all the small forces $dF$ for the small slices $dy$ between $y = 0$ m and $y = d$. This summation is expressed as the integral

$$F_{\text{total}} = \int_{\text{all slices}} dF = \int_0^d dF = \rho gw\int_0^d (d-y)dy = \rho gw\left(yd - \frac{1}{2}y^2\right)\Bigg|_0^d = \frac{1}{2}\rho gwd^2$$

**(b)** The total force is

$$F_{\text{total}} = \tfrac{1}{2}(1000\ \text{kg/m}^3)(9.8\ \text{m/s}^2)(100\ \text{m})(60\ \text{m})^2 = 1.76\times10^9\ \text{N}$$

**15.49. Visualize:**

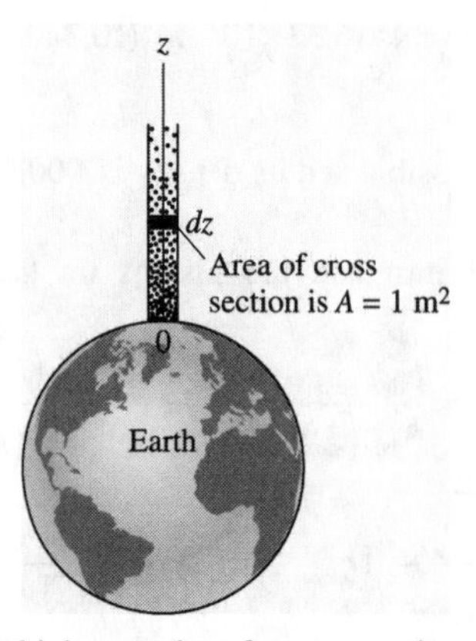


The figure shows a small column of air of thickness $dz$, of cross-sectional area $A = 1\ \text{m}^2$, and of density $\rho(z)$. The column is at a height $z$ above the surface of the earth.

**Solve:** **(a)** The atmospheric pressure at sea level is $1.013\times10^5$ Pa. That is, the weight of the air column with a 1 m$^2$ cross section is $1.013\times10^5$ N. Consider the weight of a 1 m$^2$ slice of thickness $dz$ at a height $z$. This slice has volume $dV = Adz =$ $(1\ \text{m}^2)dz$, so its weight is $dw = (\rho dV)g = \rho g(1\ \text{m}^2)dz = \rho_0 e^{-z/z_0} g(1\ \text{m}^2)dz$. The total weight of the 1 m$^2$ column is found by adding all the $dw$. Integrating from $z = 0$ to $z = \infty$,

$$\begin{aligned} w &= \int_0^\infty \rho_0 g(1\ \text{m}^2)e^{-z/z_0}dz \\ &= \left(-\rho_0 g(1\ \text{m}^2)z_0\right)\left[e^{-z/z_0}\right]_0^\infty \\ &= \rho_0 g(1\ \text{m}^2)z_0 \end{aligned}$$

Because $w = 101{,}300\ \text{N} = \rho_0 g(1\ \text{m}^2)z_0$,

$$z_0 = \frac{101{,}300\ \text{N}}{(1.28\ \text{kg/m}^3)(9.8\ \text{m/s}^2)(1.0\ \text{m}^2)} = 8.08\times10^3\ \text{m}$$

**(b)** Using the density at sea level from Table 15.1,

$$\rho = (1.28\ \text{kg/m}^3)e^{-z/(8.08\times10^3\ \text{m})} = (1.28\ \text{kg/m}^3)e^{-1600\ \text{m}/(8.08\times10^3\ \text{m})} = 1.05\ \text{kg/m}^3$$

This is 82% of $\rho_0$.

**15.55. Model:** The buoyant force is determined by Archimedes' principle. The spring is ideal.
**Visualize:**

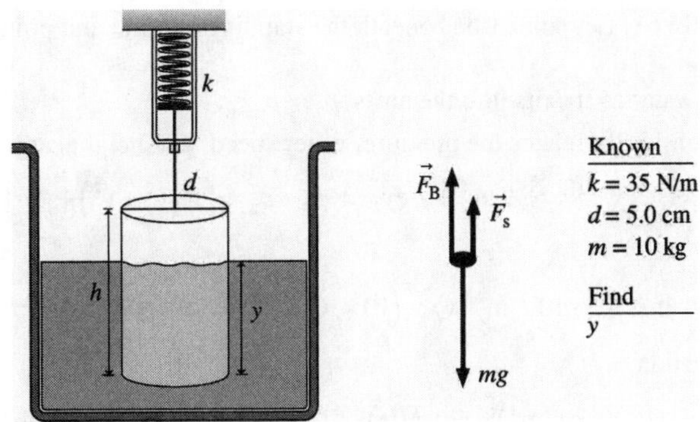


**Solve:** The spring is stretched by the same amount that the cylinder is submerged. The buoyant force and spring force balance the gravitational force on the cylinder.

$$\sum F_y = F_B + F_S - mg = 0 \text{ N}$$
$$\Rightarrow p_w Ayg + ky = mg$$
$$y = \frac{mg}{p_w Ag + k} = \frac{(1.0 \text{ kg})(9.8 \text{ m/s}^2)}{(1000 \text{ kg/m}^3)\pi(0.025 \text{ m})^2(9.8 \text{ m/s}^2) + 35 \text{ N/m}}$$
$$= 0.181 \text{ m} = 18.1 \text{ cm}$$

**Assess:** This is difficult to assess because we don't know the height $h$ of the cylinder and can't calculate it without the density of the metal material.

**15.57. Model:** The buoyant force on the can is given by Archimedes' principle.
**Visualize:**

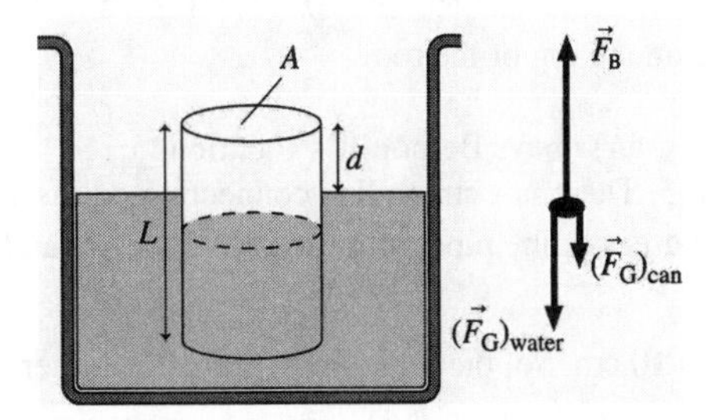


The length of the can above the water level is $d$, the length of the can is $L$, and the cross-sectional area of the can is $A$.
**Solve:** The can is in static equilibrium, so

$$\sum F_y = F_B - (F_G)_{can} - (F_G)_{water} = 0 \text{ N} \Rightarrow \rho_{water} A(L-d)g = (0.020 \text{ kg})g + m_{water} g$$

The mass of the water in the can is

$$m_{water} = \rho_{water}\left(\frac{V_{can}}{2}\right) = (1000 \text{ kg/m}^3)\frac{355\times10^{-6} \text{ m}^3}{2} = 0.1775 \text{ kg}$$
$$\Rightarrow \rho_{water} A(L-d) = 0.020 \text{ kg} + 0.1775 \text{ kg} = 0.1975 \text{ kg} \Rightarrow d - L = -\frac{0.1975 \text{ kg}}{\rho_{water} A} = 0.0654 \text{ m}$$

Because $V_{can} = \pi(0.031 \text{ m})^2 L = 355\times10^{-6} \text{ m}^3$, $L = 0.1176$ m. Using this value of $L$, we get $d = 0.0522 \text{ m} \approx 5.2$ cm.

**Assess:** $d/L = 5.22 \text{ cm}/11.76 \text{ cm} = 0.444$, thus 44.4% of the length of the can is above the water surface. This is reasonable.

**15.61. Model:** Treat the water as an ideal fluid obeying Bernoulli's equation. A streamline begins in the bigger size pipe and ends at the exit of the narrower pipe.
**Visualize:** Please see Figure P15.61. Let point 1 be beneath the standing column and point 2 be where the water exits the pipe.
**Solve:** **(a)** The pressure of the water as it exits into the air is $p_2 = p_{atmos}$.
**(b)** Bernoulli's equation, Equation 15.28, relates the pressure, water speed, and heights at points 1 and 2:

$$p_1 + \tfrac{1}{2}\rho v_1^2 + \rho g y_1 = p_2 + \tfrac{1}{2}\rho v_2^2 + \rho g y_2 \Rightarrow p_1 - p_2 = \tfrac{1}{2}\rho\left(v_2^2 - v_1^2\right) + \rho g\left(y_2 - y_1\right)$$

From the continuity equation,

$$v_1 A_1 = v_2 A_2 = (4 \text{ m/s})\left(5\times10^{-4} \text{ m}^2\right) \Rightarrow v_1\left(10\times10^{-4} \text{ m}^2\right) = 20\times10^{-4} \text{ m}^3/\text{s} \Rightarrow v_1 = 2.0 \text{ m/s}$$

Substituting into Bernoulli's equation,

$$p_1 - p_2 = p_1 - p_{atmos} = \tfrac{1}{2}\left(1000 \text{ kg/m}^3\right)\left[(4.0 \text{ m/s})^2 - (2.0 \text{ m/s})^2\right] + \left(1000 \text{ kg/m}^3\right)(9.8 \text{ m/s})(4.0 \text{ m})$$

$$= 6000 \text{ Pa} + 39{,}200 \text{ Pa} = 45 \text{ kPa}$$

But $p_1 - p_2 = \rho g h$, where *h* is the height of the standing water column. Thus

$$h = \frac{45\times10^3 \text{ Pa}}{\left(1000 \text{ kg/m}^3\right)\left(9.8 \text{ m/s}^2\right)} = 4.6 \text{ m}$$

**15.63. Model:** Treat the air as an ideal fluid obeying Bernoulli's equation.
**Solve:** **(a)** The pressure above the roof is lower due to the higher velocity of the air.
**(b)** Bernoulli's equation, with $y_{inside} \approx y_{outside}$, is

$$p_{inside} = p_{outside} + \tfrac{1}{2}\rho_{air} v^2 \Rightarrow \Delta p = \frac{1}{2}\rho_{air} v^2 = \frac{1}{2}\left(1.28 \text{ kg/m}^3\right)\left(\frac{130\times1000 \text{ m}}{3600 \text{ s}}\right)^2 = 835 \text{ Pa}$$

The pressure difference is 0.83 kPa
**(c)** The force on the roof is $(\Delta p)A = (835 \text{ Pa})(6.0 \text{ m}\times15.0 \text{ m}) = 7.5\times10^4 \text{ N}$. The roof will blow up, because pressure inside the house is greater than pressure on the top of the roof.

**15.65. Model:** The ideal fluid (that is, air) obeys Bernoulli's equation.
**Visualize:** Please refer to Figure P15.65. There is a streamline connecting points 1 and 2. The air speeds at points 1 and 2 are $v_1$ and $v_2$, and the cross-sectional areas of the pipes at these points are $A_1$ and $A_2$. Points 1 and 2 are at the same height, so $y_1 = y_2$.
**Solve:** **(a)** The height of the mercury is 10 cm. So, the pressure at point 2 is larger than at point 1 by

$$\rho_{Hg} g(0.10 \text{ m}) = \left(13{,}600 \text{ kg/m}^3\right)\left(9.8 \text{ m/s}^2\right)(0.10 \text{ m}) = 13{,}328 \text{ Pa} \Rightarrow p_2 = p_1 + 13{,}328 \text{ Pa}$$

Using Bernoulli's equation,

$$p_1 + \tfrac{1}{2}\rho_{air} v_1^2 + \rho_{air} g y_1 = p_2 + \tfrac{1}{2}\rho_{air} v_2^2 + \rho_{air} g y_2 \Rightarrow p_2 - p_1 = \tfrac{1}{2}\rho_{air}\left(v_1^2 - v_2^2\right)$$

$$\Rightarrow v_1^2 - v_2^2 = \frac{2(p_2 - p_1)}{\rho_{air}} = \frac{2(13{,}328 \text{ Pa})}{\left(1.28 \text{ kg/m}^3\right)} = 20{,}825 \text{ m}^2/\text{s}^2$$

From the continuity equation, we can obtain another equation connecting $v_1$ and $v_2$:

$$A_1 v_1 = A_2 v_2 \Rightarrow v_1 = \frac{A_2}{A_1} v_2 = \frac{\pi(0.005 \text{ m})^2}{\pi(0.001 \text{ m})^2} v_2 = 25\, v_2$$

Substituting $v_1 = 25 v_2$ in the Bernoulli equation, we get

$$(25\, v_2)^2 - v_2^2 = 20{,}825 \text{ m}^2/\text{s}^2 \Rightarrow v_2 = 5.78 \text{ m/s}$$

Thus $v_2 = 5.8$ m/s and $v_1 = 25 v_2 = 144$ m/s.
**(b)** The volume flow rate $A_2 v_2 = \pi(0.0050 \text{ m})^2(5.78 \text{ m/s}) = 4.5\times10^{-3} \text{ m}^3/\text{s}$.

**15.67. Model:** Treat water as an ideal fluid that obeys Bernoulli's equation. There is a streamline connecting the top of the tank with the hole.

**Visualize:** Please refer to Figure P15.67. We placed the origin of the coordinate system at the bottom of the tank so that the top of the tank (point 1) is at a height of $h+1.0\text{ m}$ and the hole (point 2) is at a height *h*. Both points 1 and 2 are at atmospheric pressure.

**Solve:** **(a)** Bernoulli's equation connecting points 1 and 2 is

$$p_1+\tfrac{1}{2}\rho v_1^2+\rho g y_1=p_2+\tfrac{1}{2}\rho v_2^2+\rho g y_2$$
$$\Rightarrow p_{\text{atmos}}+\tfrac{1}{2}\rho v_1^2+\rho g(h+1.0\text{ m})=p_{\text{atmos}}+\tfrac{1}{2}\rho v_2^2+\rho g h$$
$$\Rightarrow v_2^2-v_1^2=2g(1.0\text{ m})=19.6\text{ m}^2/\text{s}^2$$

Using the continuity equation $A_1v_1=A_2v_2$,

$$v_1=\left(\frac{A_2}{A_1}\right)v_2=\frac{\pi(2.0\times10^{-3}\text{ m})^2}{\pi(1.0\text{ m})^2}v_2=\frac{v_2}{250{,}000}$$

Because $v_1 \ll v_2$, we can simply put $v_1\approx 0$ m/s. Bernoulli's equation thus simplifies to

$$v_2^2=19.6\text{ m}^2/\text{s}^2\Rightarrow v_2=4.43\text{ m/s}$$

Therefore, the volume flow rate through the hole is

$$Q=A_2v_2=\pi(2.0\times10^{-3}\text{ m})^2(4.43\text{ m/s})=5.56\times10^{-5}\text{ m}^3/\text{s}=3.3\text{ L/min}$$

**(b)** The rate at which the water level will drop is

$$v_1=\frac{v_2}{250{,}000}=\frac{4.43\text{ m/s}}{250{,}000}=1.77\times10^{-2}\text{ mm/s}=1.06\text{ mm/min}$$

**Assess:** Because the hole through which water flows out of the tank has a diameter of only 4.0 mm, a drop in the water level at the rate of 1.06 mm/min is reasonable.

# 16

## A MACROSCOPIC DESCRIPTION OF MATTER

**16.3. Solve:** The volume of the aluminum cube is $10^{-3}$ m$^3$ and its mass is

$$m_{\text{Al}} = \rho_{\text{Al}} V_{\text{Al}} = \left(2700 \text{ kg/m}^3\right)\left(1.0\times10^{-3} \text{ m}^3\right) = 2.7 \text{ kg}$$

The volume of the copper sphere with this mass is

$$V_{\text{Cu}} = \frac{4\pi}{3}\left(r_{\text{Cu}}\right)^3 = \frac{m_{\text{Cu}}}{\rho_{\text{Cu}}} = \frac{2.7 \text{ kg}}{8920 \text{ kg/m}^3} = 3.027\times10^{-4} \text{ m}^3$$

$$\Rightarrow r_{\text{Cu}} = \left[\frac{3\left(3.027\times10^{-4} \text{ m}^3\right)}{4\pi}\right]^{1/3} = 0.042 \text{ m}$$

The diameter of the copper sphere is 0.0833 m = 8.33 cm.
**Assess:** The diameter of the sphere is a little less than the length of the cube, and this is reasonable considering the density of copper is greater than the density of aluminum.

**16.7. Solve:** **(a)** The number density is defined as $N/V$, where $N$ is the number of particles occupying a volume $V$. Because Al has a mass density of 2700 kg/m$^3$, a volume of 1 m$^3$ has a mass of 2700 kg. We also know that the molar mass of Al is 27 g/mol or 0.027 kg/mol. So, the number of moles in a mass of 2700 kg is

$$n = (2700 \text{ kg})\left(\frac{1 \text{ mol}}{0.027 \text{ kg}}\right) = 1.00\times10^5 \text{ mol}$$

The number of Al atoms in $1.00 \times 10^5$ mols is

$$N = nN_{\text{A}} = \left(1.00\times10^5 \text{ mol}\right)\left(6.02\times10^{23} \text{ atoms/mol}\right) = 6.02\times10^{28} \text{ atoms}$$

Thus, the number density is

$$\frac{N}{V} = \frac{6.02\times10^{28} \text{atoms}}{1 \text{ m}^3} = 6.02\times10^{28} \text{ atoms/m}^3$$

**(b)** Pb has a mass of 11,300 kg in a volume of 1 m$^3$. Since the atomic mass number of Pb is 207, the number of moles in 11,300 kg is

$$n = (11{,}300 \text{ kg})\left(\frac{1 \text{ mole}}{0.207 \text{ kg}}\right)$$

The number of Pb atoms is thus $N = nN_{\text{A}}$, and hence the number density is

$$\frac{N}{V} = \frac{nN_{\text{A}}}{V} = \left(\frac{11{,}300 \text{ kg}}{0.207 \text{ kg}}\right)\left(6.02\times10^{23} \frac{\text{atoms}}{\text{mol}}\right)\left(\frac{1 \text{ mol}}{1 \text{ m}^3}\right) = 3.28\times10^{28} \frac{\text{atoms}}{\text{m}^3}$$

**Assess:** We expected to get very large numbers like this.

**16.13. Model:** A temperature scale is a linear scale.

**Solve:** **(a)** We need a conversion formula for °C to °Z, analogous to the conversion of °C to °F. Since temperature scales are linear, $T_C = aT_Z + b$, where $a$ and $b$ are constants to be determined. We know the boiling point of liquid nitrogen is 0°Z and −196°C. Similarly, the melting point of iron is 1000°Z and 1538°C. Thus

$$-196 = 0a + b$$
$$1538 = 1000a + b$$

From the first, $b = -196°$. Then from the second, $a = (1538+196)/1000 = 1734/1000$. Thus the conversion is $T_C = (1734/1000)T_Z - 196°$. Since the boiling point of water is $T_C = 100°C$, its temperature in °Z is

$$T_Z = \left(\frac{1000}{1734}\right)(100° + 196°) = 171°Z$$

**(b)** A temperature $T_Z = 500°Z$ is

$$T_C = \left(\frac{1734}{1000}\right)500° - 196° = 671°C = 944 \text{ K}$$

**16.17. Model:** Treat the gas in the sealed container as an ideal gas.

**Solve:** **(a)** From the ideal gas law equation $pV = nRT$, the volume $V$ of the container is

$$V = \frac{nRT}{p} = \frac{(2.0 \text{ mol})(8.31 \text{ J/mol K})\left[(273+30) \text{ K}\right]}{1.013\times10^5 \text{ Pa}} = 0.050 \text{ m}^3$$

Note that pressure *must* be in Pa in the ideal-gas law.

**(b)** The before-and-after relationship of an ideal gas in a sealed container (constant volume) is

$$\frac{p_1 V}{T_1} = \frac{p_2 V}{T_2} \Rightarrow p_2 = p_1 \frac{T_2}{T_1} = (1.0 \text{ atm})\frac{(273+130) \text{ K}}{(273+30) \text{ K}} = 1.3 \text{ atm}$$

Note that gas-law calculations *must* use T in kelvins.

**16.21. Model:** Treat the helium gas in the sealed cylinder as an ideal gas.

**Solve:** The volume of the cylinder is $V = \pi r^2 h = \pi(0.05 \text{ m})^2(0.30 \text{ m}) = 2.356\times10^{-3} \text{ m}^3$. The gauge pressure of the gas is $120 \text{ psi} \times \frac{1 \text{ atm}}{14.7 \text{ psi}} \times \frac{1.013\times10^5 \text{ Pa}}{1 \text{ atm}} = 8.269\times10^5 \text{ Pa}$, so the absolute pressure of the gas is $8.269\times10^5 \text{ Pa} + 1.013\times10^5 \text{ Pa} = 9.282\times10^5 \text{ Pa}$. The temperature of the gas is $T = (273+20) \text{ K} = 293 \text{ K}$. The number of moles of the gas in the cylinder is

$$n = \frac{pV}{RT} = \frac{(9.282\times10^5 \text{ Pa})(2.356\times10^{-3} \text{ m}^3)}{(8.31 \text{ J/mol K})(293 \text{ K})} = 0.898 \text{ mol}$$

**(a)** The number of atoms is

$$N = nN_A = (0.898 \text{ mol})(6.02\times10^{23} \text{ mol}^{-1}) = 5.41\times10^{23} \text{ atoms} \approx 5.4\times10^{23} \text{ atoms}$$

**(b)** The mass of the helium is

$$M = nM_{\text{mol}} = (0.898 \text{ mol})(4 \text{ g/mol}) = 3.59 \text{ g} = 3.59\times10^{-3} \text{ kg} \approx 3.6\times10^{-3} \text{ kg}$$

**(c)** The number density is

$$\frac{N}{V} = \frac{5.41\times10^{23} \text{ atoms}}{2.356\times10^{-3} \text{ m}^3} = 2.3\times10^{26} \text{ atoms/m}^3$$

**(d)** The mass density is

$$\rho = \frac{M}{V} = \frac{3.59\times10^{-3} \text{ kg}}{2.356\times10^{-3} \text{ m}^3} = 1.5 \text{ kg/m}^3$$

**16.27. Model:** In an isochoric process, the volume of the container stays unchanged. Argon gas in the container is assumed to be an ideal gas.

**Solve:** **(a)** The container has only argon inside with $n = 0.1$ mol, $V_1 = 50\text{ cm}^3 = 50\times10^{-6}\text{ m}^3$, and $T_1 = 20°\text{C} = 293$ K. This produces a pressure

$$p_1 = \frac{nRT}{V_1} = \frac{(0.1\text{ mol})(8.31\text{ J/mol K})(293\text{ K})}{50\times10^{-6}\text{ m}^3} = 4.87\times10^6\text{ Pa} = 4870\text{ kPa} \approx 4900\text{ kPa}$$

An ideal gas process has $p_2V_2/T_2 = p_1V_1/T_1$. Isochoric heating to a final temperature $T_2 = 300°\text{C} = 573\text{ K}$ has $V_2 = V_1$, so the final pressure is

$$p_2 = \frac{V_1}{V_2}\frac{T_2}{T_1}p_1 = 1\times\frac{573}{293}\times4870\text{ kPa} = 9520\text{ kPa} \approx 9500\text{ kPa}$$

Note that it is essential to express temperatures in kelvins.

**(b)**

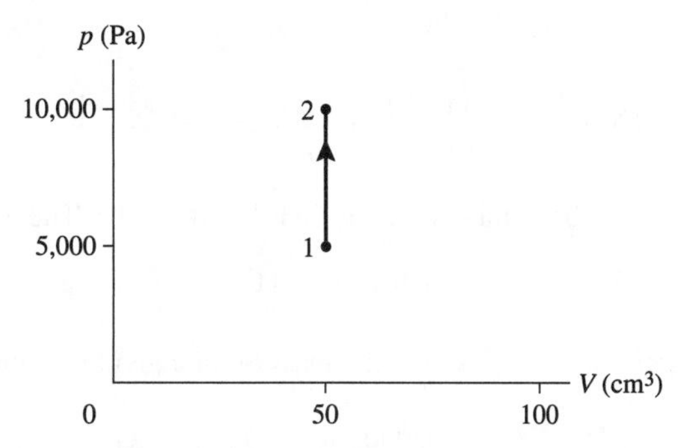


**Assess:** All isochoric processes will be a straight vertical line on a $pV$ diagram.

**16.29. Model:** In an isothermal expansion, the temperature stays the same. The argon gas in the container is assumed to be an ideal gas.

**Solve:** **(a)** The container has only argon inside with $n = 0.1$ mol, $V_1 = 50\text{ cm}^3 = 50\times10^{-6}\text{ m}^3$, and $T_1 = 20°\text{C} = 293$ K. This produces a pressure

$$p_1 = \frac{nRT_1}{V_1} = \frac{(0.1\text{ mol})(8.31\text{ J/mol K})(293\text{ K})}{50\times10^{-6}\text{ Pa}} = 4.87\times10^6\text{ Pa} = 12.02\text{ atm} \approx 12\text{ atm}$$

An ideal-gas process obeys $p_2V_2/T_2 = p_1V_1/T_1$. Isothermal expansion to $V_2 = 200\text{ cm}^3$ gives a final pressure

$$p_2 = \frac{T_2}{T_1}\frac{V_1}{V_2}p_1 = 1\times\frac{200}{50}\times12.02\text{ atm} = 48\text{ atm}$$

**(b)**

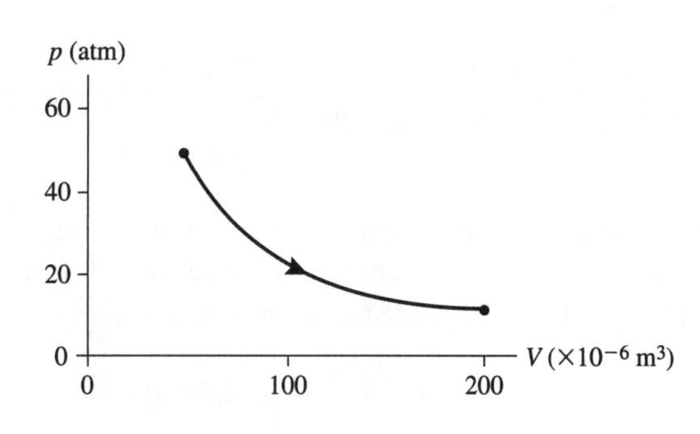

**16.33. Visualize:**

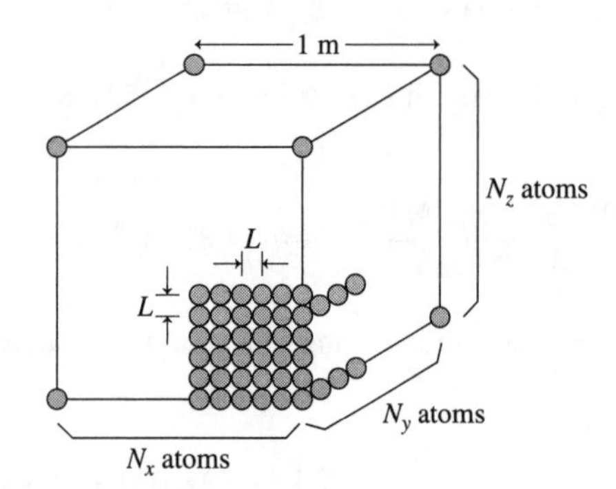


**Solve:** Suppose we have a $1\ \text{m}\times 1\ \text{m}\times 1\ \text{m}$ block of copper of mass $M$ containing $N$ atoms. The atoms are spaced a distance $L$ apart along all three axes of the cube. There are $N_x$ atoms along the $x$-edge of the cube, $N_y$ atoms along the $y$-edge, and $N_z$ atoms along the $z$-edge. The total number of atoms is $N = N_x\ N_y\ N_z$. If $L$ is expressed in meters, then the number of atoms along the $x$-edge is $N_x = (1\ \text{m})/L$. Thus,

$$N = \frac{1\ \text{m}}{L}\times\frac{1\ \text{m}}{L}\times\frac{1\ \text{m}}{L} = \frac{1\ \text{m}^3}{L^3} \Rightarrow L = \left(\frac{1\ \text{m}^3}{N}\right)^{1/3}$$

This relates the spacing between atoms to the number of atoms in a 1-meter cube. The mass of the large cube of copper is

$$M = \rho_{\text{Cu}}V = \left(8920\ \text{kg/m}^3\right)\left(1\ \text{m}^3\right) = 8920\ \text{kg}$$

But $M = mN$, where $m = 64\ \text{u} = 64\times\left(1.661\times10^{-27}\ \text{kg}\right)$ is the mass of an individual copper atom. Thus,

$$N = \frac{M}{m} = \frac{8920\ \text{kg}}{64\times\left(1.661\times10^{-27}\ \text{kg}\right)} = 8.39\times10^{28}\ \text{atoms}$$

$$\Rightarrow L = \left(\frac{1\ \text{m}^3}{8.39\times10^{28}}\right)^{1/3} = 2.28\times10^{-10}\ \text{m}\ =\ 0.228\ \text{nm}$$

**Assess:** This is a reasonable interatomic spacing in a crystal lattice.

**16.39. Model:** Assume that the gas in the vacuum chamber is an ideal gas.
**Solve:** **(a)** The fraction is

$$\frac{p_{\text{vacuum chamber}}}{p_{\text{atmosphere}}} = \frac{1.0\times10^{-10}\ \text{mm of Hg}}{760\ \text{mm of Hg}} = 1.3\times10^{-13}$$

**(b)** The volume of the chamber $V = \pi(0.20\ \text{m})^2(0.30\ \text{m}) = 0.03770\ \text{m}^3$. From the ideal-gas equation $pV = Nk_{\text{B}}T$, the number of molecules of gas in the chamber is

$$N = \frac{pV}{k_{\text{B}}T} = \frac{\left(1.32\times10^{-13}\right)\left(1.013\times10^5\ \text{Pa}\right)\left(0.03770\ \text{m}^3\right)}{\left(1.38\times10^{-23}\ \text{J/K}\right)(293\ \text{K})} = 1.2\times10^{11}\ \text{molecules}$$

**16.47. Model:** We assume that the volume of the tire and that of the air in the tire is constant.
**Solve:** A gauge pressure of 30 psi corresponds to an absolute pressure of $(30\ \text{psi}) + (14.7\ \text{psi}) = 44.7\ \text{psi}$. Using the before-and-after relationship of an ideal gas for an isochoric (constant volume) process,

$$\frac{p_1}{T_1} = \frac{p_2}{T_2} \Rightarrow p_2 = \frac{T_2}{T_1}p_1 = \left(\frac{273+45}{273+15}\right)(44.7\ \text{psi}) = 49.4\ \text{psi}$$

Your tire gauge will read a gauge pressure $p_{\text{g}} = 49.4\ \text{psi} - 14.7\ \text{psi} = 34.7\ \text{psi.} \approx 35\ \text{psi.}$

**16.51. Model:** The air in the closed section of the U-tube is an ideal gas.
**Visualize:** The length of the tube is $l = 1.0$ m and its cross-sectional area is $A$.
**Solve:** Initially, the pressure of the air in the tube is $p_1 = p_{\text{atmos}}$ and its volume is $V_1 = Al$. After the mercury is poured in, compressing the air, the air-pressure force supports the weight of the mercury. Thus the compressed pressure equals the pressure at the bottom of the column: $p_2 = p_{\text{atmos}} + \rho gL$. The volume of the compressed air is $V_2 = A(l - L)$. Because the mercury is poured in slowly, we will assume that the gas remains in thermal equilibrium with the surrounding air, so $T_2 = T_1$. In an isothermal process, pressure and volume are related by

$$p_1V_1 = p_{\text{atmos}}Al = p_2V_2 = (p_{\text{atmos}} + \rho gL)A(l - L)$$

Canceling the $A$, multiplying through, and solving for $L$ gives

$$L = l - \frac{p_{\text{atmos}}}{\rho g} = 1.00 \text{ m} - \frac{101{,}300 \text{ Pa}}{(13{,}600 \text{ kg/m}^3)(9.8 \text{ m/s}^2)} = 0.240 \text{ m} = 24.0 \text{ cm}$$

**16.55. Model:** Assume that the gas is an ideal gas.

**Solve:**

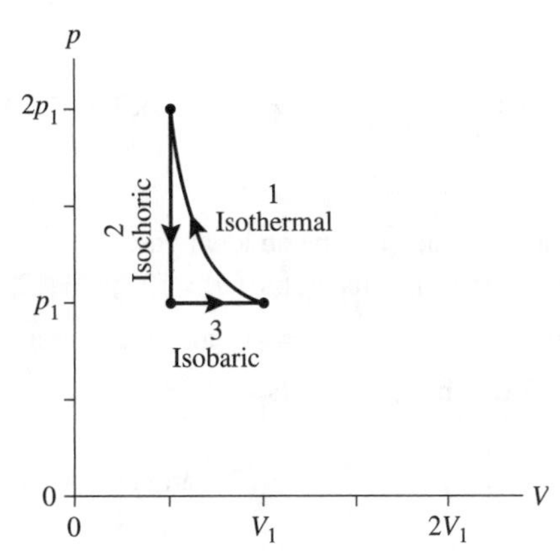


**Assess:** For the isothermal process, the pressure must double as the volume is halved. This is because $p$ is proportional to $1/V$ for isothermal processes.

**16.57. Model:** Assume the nitrogen gas is an ideal gas.
**Solve:** **(a)** The number of moles of nitrogen is

$$n = \frac{M}{M_{\text{mol}}} = \frac{1 \text{ g}}{28 \text{ g/mol}} = \left(\frac{1}{28}\right) \text{mol}$$

Using the ideal-gas equation,

$$p_1 = \frac{nRT_1}{V_1} = \frac{(1/28 \text{ mol})(8.31 \text{ J/mol K})(298 \text{ K})}{\left(100 \times 10^{-6} \text{ m}^3\right)} = 8.84 \times 10^5 \text{ Pa} = 884 \text{ kPa} \approx 880 \text{ kPa}$$

**(b)** For the process from state 1 to state 3:

$$\frac{p_1V_1}{T_1} = \frac{p_3V_3}{T_3} \Rightarrow T_3 = T_1\frac{p_3}{p_1}\frac{V_3}{V_1} = (298 \text{ K})\left(\frac{1.5p_1}{p_1}\right)\left(\frac{50 \text{ cm}^3}{100 \text{ cm}^3}\right) = 223.5 \text{ K} \approx -49°\text{C}$$

For the process from state 3 to state 2:

$$\frac{p_2V_2}{T_2} = \frac{p_3V_3}{T_3} \Rightarrow T_2 = T_3\left(\frac{p_2}{p_3}\right)\left(\frac{V_2}{V_3}\right) = (223.5 \text{ K})\left(\frac{2.0p_1}{1.5p_1}\right)\left(\frac{100 \text{ cm}^3}{50 \text{ cm}^3}\right) = 596 \text{ K} = 323°\text{C}$$

For the process from state 1 to state 4:

$$\frac{p_4V_4}{T_4}=\frac{p_1V_1}{T_1}\Rightarrow T_4=T_1\frac{p_4}{p_1}\frac{V_4}{V_1}=(298\text{ K})\left(\frac{1.5p_1}{p_1}\right)\left(\frac{150\text{ cm}^3}{100\text{ cm}^3}\right)=670.5\text{ K}\approx 398°\text{C}$$

**16.59. Model:** Assume the gas is an ideal gas.
**Solve:** (a) We can find the temperatures directly from the ideal-gas law after we convert all quantities to SI units:

$$T_1=\frac{p_1V_1}{nR}=\frac{(3.0\text{ atm}\times 101{,}300\text{ Pa/atm})(1000\text{ cm}^3\times 10^{-6}\text{ m}^3/\text{cm}^3)}{(0.10\text{ mol})(8.31\text{ J/mol K})}=366\text{ K}=93°\text{C}$$

$$T_2=\frac{p_2V_2}{nR}=\frac{(1.0\text{ atm}\times 101{,}300\text{ Pa/atm})(3000\text{ cm}^3\times 10^{-6}\text{ m}^3/\text{cm}^3)}{(0.10\text{ mol})(8.31\text{ J/mol K})}=366\text{ K}=93°\text{C}$$

**(b)** $T_2=T_1$, so this is an isothermal process.

**(c)** A constant volume process has $V_3=V_2$. Because $p_1=3p_2$, restoring the pressure to its original value means that $p_3=3p_2$. From the ideal-gas law,

$$\frac{p_3V_3}{T_3}=\frac{p_2V_2}{T_2}\Rightarrow T_3=\left(\frac{p_3}{p_2}\right)\left(\frac{V_3}{V_2}\right)T_2=3\times 1\times T_2=3\times 366\text{ K}=1098\text{ K}=825°\text{C}$$

**16.63. Model:** The gas in the container is assumed to be an ideal gas.
**Solve:** (a) The gas starts at pressure $p_1=2.0$ atm, temperature $T_1=127°\text{C}=(127+273)\text{ K}=400\text{ K}$ and volume $V_1$. It is first compressed at a constant temperature $T_2=T_1$ until $V_2=\frac{1}{2}V_1$ and the pressure is $p_2$. It is then further compressed at constant pressure $p_3=p_2$ until $V_3=\frac{1}{2}V_2$. From the ideal-gas law,

$$\frac{p_2V_2}{T_2}=\frac{p_1V_1}{T_1}\Rightarrow p_2=p_1\frac{V_1}{V_2}\frac{T_2}{T_1}=(2.0\text{ atm})\frac{V_1}{\frac{1}{2}V_1}\times 1=4.0\text{ atm}$$

Note that $T_2=T_1=400$ K. Using the ideal-gas law once again,

$$\frac{p_3V_3}{T_3}=\frac{p_2V_2}{T_2}\Rightarrow T_3=T_2\frac{V_3}{V_2}\frac{p_3}{p_2}=(400\text{ K})\frac{\frac{1}{2}V_2}{V_2}\times 1=200\text{ K}=-73°\text{C}$$

The final pressure and temperature are 4.0 atm and –73°C.

**(b)**

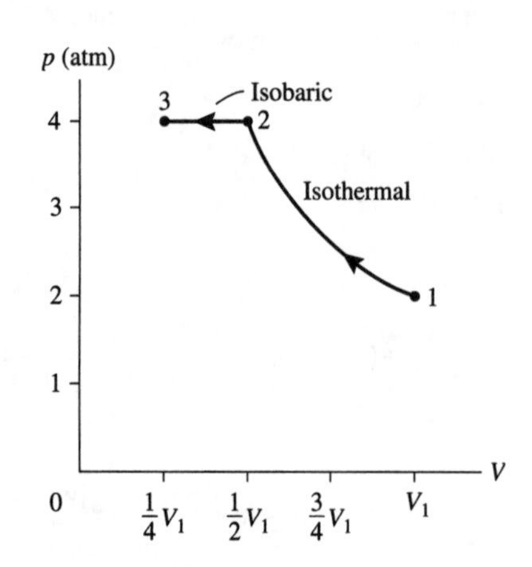

**16.65. Solve:** **(a)** A gas is compressed isothermally from a volume 300 $cm^3$ at 2 atm to a volume of 100 $cm^3$. What is the final pressure?

**(b)**

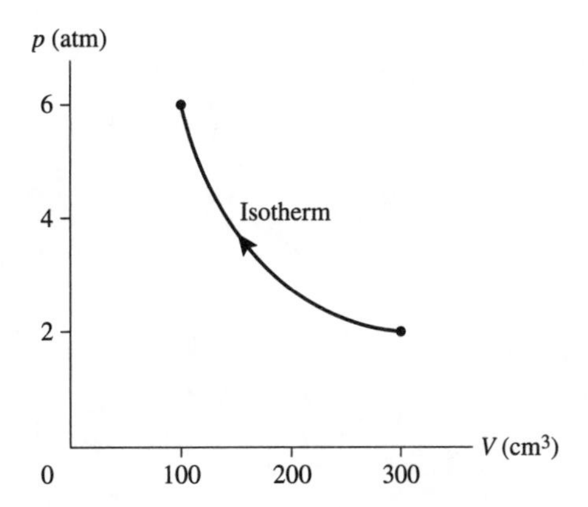


**(c)** The final pressure is $p_2 = 6\text{ atm}$.

**16.67. Solve:** **(a)** A gas expands at constant pressure from 200 $cm^3$ at 50°C until the temperature is 400°C. What is the final volume?

**(b)**

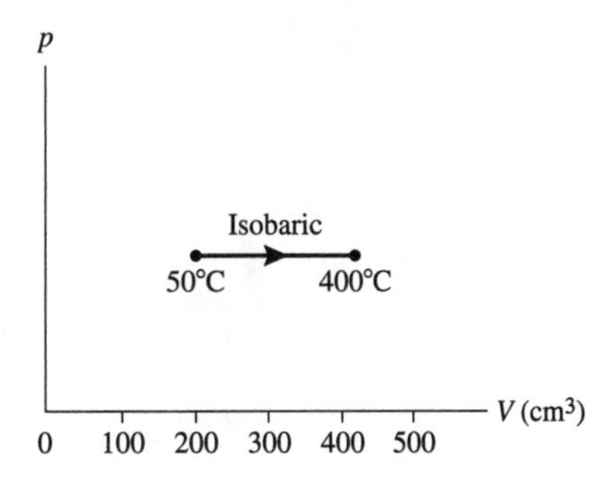


**(c)** The final volume is $V_2 = 417\text{ cm}^3$.

# 17 WORK, HEAT, AND THE FIRST LAW OF THERMODYNAMICS

**17.1. Model:** Assume the gas is ideal. The work done on a gas is the negative of the area under the $pV$ curve.
**Visualize:** The gas is compressing, so we expect the work to be positive.
**Solve:** The work done on the gas is

$$W = -\int p\,dV = -(\text{area under the } pV \text{ curve})$$
$$= -\left(-(200\ \text{cm}^3)(200\ \text{kPa})\right) = (200\times10^{-6}\ \text{m}^3)(2.0\times10^5\ \text{Pa}) = 40\ \text{J}$$

**Assess:** The area under the curve is negative because the integration direction is to the left. Thus, the environment does positive work on the gas to compress it.

**17.3. Model:** Assume the gas is ideal. The work done on a gas is the negative of the area under the $pV$ curve.
**Solve:** The work done on gas in an isobaric process is

$$W = -p\Delta V = -p(V_f - V_i)$$

Substituting into this equation,

$$80\ \text{J} = -(200\times10^3\ \text{Pa})(V_i - 3V_i) \Rightarrow V_i = 2.0\times10^{-4}\ \text{m}^3 = 200\ \text{cm}^3$$

**Assess:** The work done to compress a gas is positive.

**17.5. Visualize:**

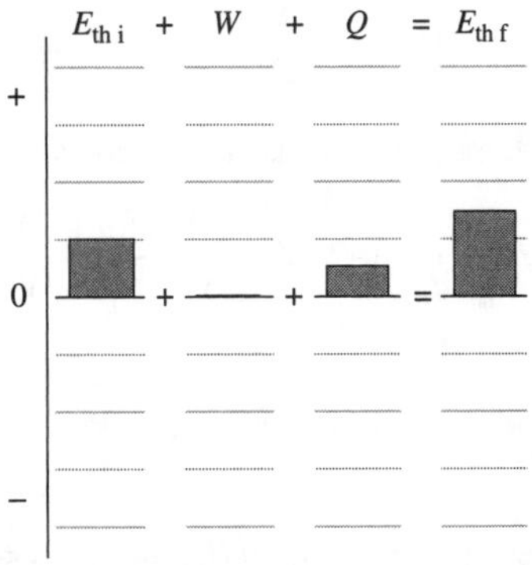

$E_{th\,i}$ + $W$ + $Q$ = $E_{th\,f}$

**Solve:** Because $W = -\int p\,dV$ and this is an isochoric process, $W = 0$ J. The final point is on a higher isotherm than the initial point, so $T_f > T_i$. Heat energy is thus transferred into the gas $(Q > 0)$ and the thermal energy of the gas increases $(E_{th\,f} > E_{th\,i})$ as the temperature increases.

**17.9. Solve:** The first law of thermodynamics is

$$\Delta E_{th} = W + Q \Rightarrow -200\ \text{J} = 500\ \text{J} + Q = Q \Rightarrow Q = -700\ \text{J}$$

The negative sign means a transfer of energy from the system to the environment.

**Assess:** Because $W > 0$ means a transfer of energy into the system, $Q$ must be less than zero and larger in magnitude than $W$ so that $E_{\text{th f}} < E_{\text{th i}}$.

**17.13. Model:** Heating the mercury at its boiling point changes its thermal energy without a change in temperature.

**Solve:** The mass of the mercury is $M = 20\text{ g} = 2.0\times10^{-2}$ kg, the specific heat $c_{\text{mercury}} = 140$ J/kg K, the boiling point $T_b = 357°\text{C}$, and the heat of vaporization $L_v = 2.96\times10^5$ J/kg. The heat required for the mercury to change to the vapor phase is the sum of two steps. The first step is

$$Q_1 = Mc_{\text{mercury}}\Delta T = (2.0\times10^{-2}\text{ kg})(140\text{ J/kg K})(357°\text{C} - 20°\text{C}) = 940\text{ J}$$

The second step is

$$Q_2 = ML_V = (2.0\times10^{-2}\text{ kg})(2.96\times10^5\text{ J/kg}) = 5920\text{ J}$$

The total heat needed is 6860 J.

**17.17. Model:** We have a thermal interaction between the copper pellets and the water.

**Solve:** The conservation of energy equation $Q_c + W_w = 0$ is

$$M_c c_c(T_f - 300°\text{C}) + M_w c_w(T_f - 20°\text{C}) = 0\text{ J}$$

Solving this equation for the final temperature $T_f$ gives

$$T_f = \frac{M_c c_c(300°\text{C}) + M_w c_w(20°\text{C})}{M_c c_c + M_w c_w}$$

$$= \frac{(0.030\text{ kg})(385\text{ J/kg K})(300°\text{C}) + (0.10\text{ kg})(4190\text{ J/kg K})(20°\text{C})}{(0.030\text{ kg})(385\text{ J/kg K}) + (0.10\text{ kg})(4190\text{ J/kg K})} = 28°\text{C}$$

The final temperature of the water and the copper is 28°C.

**17.27. Model:** The $O_2$ gas has $\gamma = 1.40$ and is an ideal gas.

**Solve:** **(a)** For an adiabatic process, $pV^\gamma$ remains a constant. That is,

$$p_i V_i^\gamma = p_f V_f^\gamma \Rightarrow p_f = p_i\left(\frac{V_i}{V_f}\right)^\gamma = (3.0\text{ atm})\left(\frac{V_i}{2V_i}\right)^{1.40} = (3.0\text{ atm})\left(\frac{1}{2}\right)^{1.40} = 1.14\text{ atm} \approx 1.1\text{ atm}$$

**(b)** Using the ideal-gas law, the final temperature of the gas is calculated as follows:

$$\frac{p_i V_i}{T_i} = \frac{p_f V_f}{T_f} \Rightarrow T_f = T_i \frac{p_f}{p_i}\frac{V_f}{V_i} = (423\text{ K})\left(\frac{1.14\text{ atm}}{3.0\text{ atm}}\right)\left(\frac{2V_i}{V_i}\right) = 321.5\text{ K} \approx 48°\text{C}$$

**17.33. Solve:** The area of the garden pond is $A = \pi(2.5\text{ m})^2 = 19.635\text{ m}^2$ and its volume is $V = A(0.30\text{ m}) = 5.891\text{ m}^3$. The mass of water in the pond is

$$M = \rho V = (1000\text{ kg/m}^3)(19.635\text{ m}^3) = 5891\text{ kg}$$

The water absorbs all the solar power which is

$$(400\text{ W/m}^2)(19.635\text{ m}^2) = 7854\text{ W}$$

This power is used to raise the temperature of the water. That is,

$$Q = (7854\text{ W})\Delta t = Mc_{\text{water}}\Delta T = (5891\text{ kg})(4190\text{ J/kg K})(10\text{ K}) \Rightarrow \Delta t = 31{,}425\text{ s} \approx 8.7\text{ h}$$

**17.37. Model:** There are three interacting systems: aluminum, copper, and ethyl alcohol.
**Solve:** The aluminum, copper, and alcohol form a closed system, so $Q = Q_{Al} + Q_{Cu} + Q_{eth} = 0$ J. The mass of the alcohol is

$$M_{eth} = \rho V = (790\ \text{kg/m}^3)(50\times10^{-6}\ \text{m}^3) = 0.0395\ \text{kg}$$

Expressed in terms of specific heats and using the fact that $\Delta T = T_f - T_i$, the $Q = 0$ J condition is

$$M_{Al}c_{Al}\Delta T_{Al} + M_{Cu}c_{Cu}\Delta T_{Cu} + M_{eth}c_{eth}\Delta T_{eth} = 0\ \text{J}$$

Substituting into this expression,

$$(0.010\ \text{kg})(900\ \text{J/kg K})(298\ \text{K} - 473\ \text{K}) + (0.020\ \text{kg})(385\ \text{J/kg K})(298\ \text{K} - T)$$
$$+(0.0395\ \text{kg})(2400\ \text{J/kg K})(298\ \text{K} - 288\ \text{K}) = -1575\ \text{J} + (7.7\ \text{J/K})(298 - T) + 948\ \text{J} = 0\ \text{J}$$

$$\Rightarrow T = 216.6\ \text{K} = -56.4°\text{C} \approx -56°\text{C}$$

**17.41. Model:** Heating the water raises its thermal energy and its temperature.
**Solve:** A 5.0 kW heater has power $P = 5000$ W. That is, it supplies heat energy at the rate 5000 J/s. The heat supplied in time $\Delta t$ is $Q = 5000\Delta t$ J. The temperature increase is $\Delta T_C = (5/9)\Delta T_F = (5/9)(75°) = 41.67°\text{C}$. Thus

$$Q = 5000\Delta t\ \text{J} = Mc_w\Delta T = (150\ \text{kg})(4190\ \text{J/kg K})(41.67°\text{C}) \Rightarrow \Delta t = 5283\ \text{s} \approx 87\ \text{min}$$

**Assess:** A time of $\approx$1.5 hours to heat 40 gallons of water is reasonable.

**17.43. Model:** The liquefaction of the nitrogen occurs in two steps: lowering nitrogen's temperature from 20°C to −196°C, and then liquefying it at −196°C. Assume the cooling occurs at a constant pressure of 1 atm.
**Solve:** The mass of 1.0 L of liquid nitrogen is $M = \rho V = (810\ \text{kg/m}^3)(10^{-3}\ \text{m}^3) = 0.810\ \text{kg}$. This mass corresponds to

$$n = \frac{M}{M_{mol}} = \frac{810\ \text{g}}{28\ \text{g/mol}} = 28.9\ \text{mols}$$

At constant atmospheric pressure, the heat to be removed from 28.93 mols of nitrogen is

$$Q = ML_v + nC_P\Delta T$$
$$= -(0.810\ \text{kg})(1.99\times10^5\ \text{J/kg}) + (28.9\ \text{mols})(29.1\ \text{J/mol K})(77\ \text{K} - 293\ \text{K}) = -3.4\times10^5\ \text{J}$$

**17.51. Model:** This is an isobaric process.
**Visualize:**

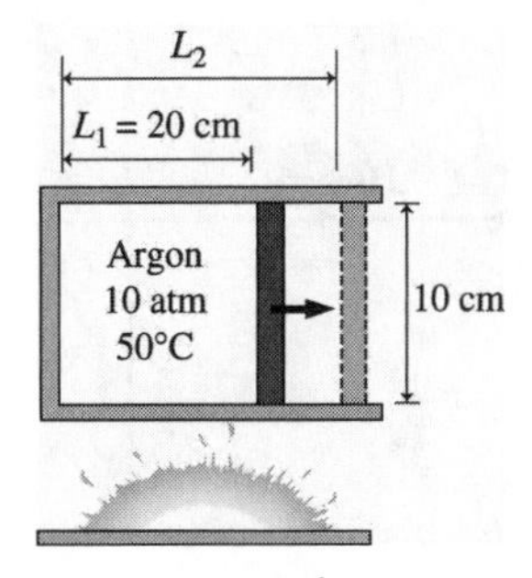


**Solve:** **(a)** The initial conditions are $p_1 = 10$ atm $= 1.013 \times 10^6$ Pa, $T_1 = 50°\text{C} = 323$ K, $V_1 = \pi r^2 L_1 = \pi(0.050\ \text{m})^2(0.20\ \text{m}) = 1.57 \times 10^{-3}\ \text{m}^3$. The gas is heated at a constant pressure, so heat and temperature change are related by $Q = nC_P\Delta T$. From the ideal gas law, the number of moles of gas is

$$n = \frac{p_1V_1}{RT_1} = \frac{(1.013\times10^6\ \text{Pa})(1.57\times10^{-3}\ \text{m}^3)}{(8.31\ \text{J/mol K})(323\ \text{K})} = 0.593\ \text{mol}$$

The temperature change due to the addition of $Q = 2500$ J of heat is thus

$$\Delta T = \frac{Q}{nC_P} = \frac{2500 \text{ J}}{(0.593 \text{ mol})(20.8 \text{ J/mol K})} = 203 \text{ K}$$

The final temperature is $T_2 = T_1 + \Delta T = 526$ K $= 253°$C.
**(b)** Noting that the volume of a cylinder is $V = \pi r^2 L$ and that $r$ doesn't change, the ideal gas relationship for an isobaric process is

$$\frac{V_2}{T_2} = \frac{V_1}{T_1} \Rightarrow \frac{L_2}{T_2} = \frac{L_1}{T_1} \Rightarrow L_2 = \frac{T_2}{T_1} L_1 = \frac{526 \text{ K}}{323 \text{ K}}(20 \text{ cm}) = 33 \text{ cm}$$

**17.55. Model:** The gas is an ideal gas and it goes through an isobaric and an isochoric process.
**Solve:** **(a)** The initial conditions are $p_1 = 3.0$ atm $= 304{,}000$ Pa and $T_1 = 293$ K. Nitrogen has a molar mass $M_{mol} =$ 28 g/mol, so 5 g of nitrogen gas has $n = M/M_{mol} = 0.1786$ mol. From this, we can find the initial volume:

$$V_1 = \frac{nRT_1}{p_1} = \frac{(0.1786 \text{ mol})(8.31 \text{ J/mol K})(293 \text{ K})}{304{,}000 \text{ Pa}} = 1.430 \times 10^{-3} \text{ m}^3 \approx 1400 \text{ cm}^3$$

The volume triples, so $V_2 = 3V_1 = 4300$ cm$^3$. The expansion is isobaric ($p_2 = p_1 = 3.0$ atm), so

$$\frac{V_2}{T_2} = \frac{V_1}{T_1} \Rightarrow T_2 = \frac{V_2}{V_1} T_1 = (3)293 \text{ K} = 879 \text{ K} = 606°\text{C}$$

**(b)** The process is isobaric, so

$$Q = nC_P \Delta T = (0.1786 \text{ mol})(29.1 \text{ J/mol K})(879 \text{ K} - 293 \text{ K}) = 3000 \text{ J}$$

**(c)** The pressure is decreased at constant volume ($V_3 = V_2 = 4290$ cm$^3$) until the original temperature is reached ($T_3 = T_1 =$ 293 K). For an isochoric process,

$$\frac{p_3}{T_3} = \frac{p_2}{T_2} \Rightarrow p_3 = \frac{T_3}{T_2} p_2 = \frac{293 \text{ K}}{879 \text{ K}}(3.0 \text{ atm}) = 1.0 \text{ atm}$$

**(d)** The process is isochoric, so

$$Q = nC_V \Delta T = (0.1786 \text{ mol})(20.8 \text{ J/mol K})(293 \text{ K} - 879 \text{ K}) = -2200 \text{ J}$$

So, 2200 J of heat was removed to decrease the pressure.
**(e)**

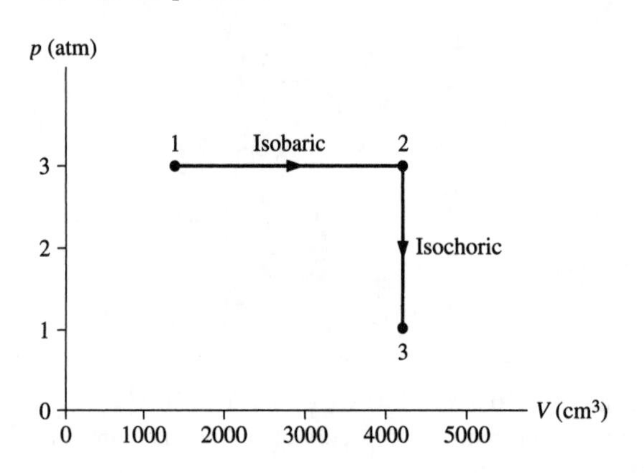


**17.57. Model:** The two processes are isochoric and isobaric.
**Solve:** Process A is isochoric which means

$$T_f/T_i = p_f/p_i \Rightarrow T_f = T_i(p_f/p_i) = T_i(1 \text{ atm}/3 \text{ atm}) = \tfrac{1}{3}T_i$$

From the ideal-gas equation,

$$T_i = \frac{p_i V_i}{nR} = \frac{(3\times1.013\times10^5 \text{ Pa})(2000\times10^{-6} \text{ m}^3)}{(0.10 \text{ mol})(8.31 \text{ J/mol K})} = 731.4 \text{ K} \Rightarrow T_f = \tfrac{1}{3}T_i = 243.8 \text{ K}$$

$$\Rightarrow T_f - T_i = -487.6 \text{ K}$$

Thus, the heat required for process A is

$$Q_A = nC_V\Delta T = (0.10 \text{ mol})(20.8 \text{ J/mol K})(-487.6 \text{ K}) = -1000 \text{ J}$$

Process B is isobaric which means

$$T_f/V_f = T_i/V_i \Rightarrow T_f = T_i(V_f/V_i) = T_i(3000 \text{ cm}^3/1000 \text{ cm}^3) = 3T_i$$

From the ideal-gas equation,

$$T_i = \frac{p_i V_i}{nR} = \frac{(2\times1.013\times10^5 \text{ Pa})(1000\times10^{-6} \text{ m}^3)}{(0.10 \text{ mol})(8.31 \text{ J/mol K})} = 243.8 \text{ K}$$

$$\Rightarrow T_f = 3T_i = 731.4 \text{ K} \Rightarrow T_f - T_i = 487.6 \text{ K}$$

Thus, heat required for process B is

$$Q_B = nC_P\Delta T = (0.10 \text{ mol})(29.1 \text{ J/mol K})(487.6 \text{ K}) = 1400 \text{ J}$$

**Assess:** Heat is transferred out of the gas in process A, but transferred into the gas in process B.

**17.61. Model:** Assume that the gas is an ideal gas. A diatomic gas has $\gamma = 1.40$.
**Solve:** **(a)** For container A,

$$V_{iA} = \frac{nRT_{Ai}}{p_{Ai}} = \frac{(0.10 \text{ mol})(8.31 \text{ J/mol K})(300 \text{ K})}{3.0\times1.013\times10^5 \text{ Pa}} = 8.20\times10^{-4} \text{ m}^3$$

For an isothermal process $p_{Af}V_{Af} = p_{Ai}V_{Ai}$. This means $T_{Af} = T_{Ai} = 300 \text{ K}$ and

$$V_{Af} = V_{Ai}(p_{Ai}/p_{Af}) = (8.20\times10^{-4} \text{ m}^3)(3.0 \text{ atm}/1.0 \text{ atm}) = 2.5\times10^{-3} \text{ m}^3$$

The gas in container B starts with the same initial volume. For an adiabatic process,

$$p_{Bf}V_{Bf}^{\gamma} = p_{Bi}V_{Bi}^{\gamma} \Rightarrow V_{Bf} = V_{Bi}\left(\frac{p_{Bi}}{p_{Bf}}\right)^{1/\gamma} = (8.20\times10^{-4}\text{m}^3)\left(\frac{3.0 \text{ atm}}{1.0 \text{ atm}}\right)^{1/1.40} = 1.8\times10^{-3} \text{ m}^3$$

The final temperature $T_{Bf}$ can now be obtained by using the ideal-gas equation:

$$T_{Bf} = T_{iB}\frac{p_{Bf}}{p_{Bi}}\frac{V_{Bf}}{V_{Bi}} = (300 \text{ K})\left(\frac{1.0 \text{ atm}}{3.0 \text{ atm}}\right)\left(\frac{1.80\times10^{-3} \text{ m}^3}{8.20\times10^{-4} \text{ m}^3}\right) = 220 \text{ K}$$

**(b)**

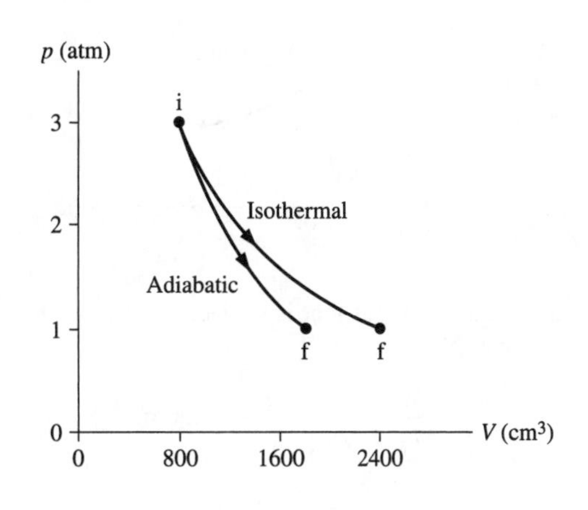

**17.67. Model:** Air is assumed to be an ideal diatomic gas that is subjected to an adiabatic process.
**Solve:** **(a)** Equation 17.40 for an adiabatic process is

$$T_f V_f^{\gamma-1} = T_i V_i^{\gamma-1} \Rightarrow \frac{V_f}{V_i} = \left(\frac{T_i}{T_f}\right)^{\frac{1}{\gamma-1}}$$

For the temperature to increase from $T_i = 20°C = 293$ K to $T_f = 1000°C = 1273$ K, the compression ratio will be

$$\frac{V_f}{V_i} = \left(\frac{293\ \text{K}}{1273\ \text{K}}\right)^{\frac{1}{1.4-1}} = 0.02542 \Rightarrow \frac{V_{max}}{V_{min}} = \frac{1}{0.02542} = 39.3$$

**(b)** From the Equation 17.39,

$$p_f V_f^{\gamma} = p_i V_i^{\gamma} \Rightarrow \frac{p_f}{p_i} = \left(\frac{V_i}{V_f}\right)^{\gamma} = (39.3)^{1.4} = 171$$

**17.69. Model:** The helium gas is assumed to be an ideal gas that is subjected to an isothermal process.
**Solve:** **(a)** The number of moles in 2.0 g of helium gas is

$$n = \frac{M}{M_{mol}} = \frac{2.0\ \text{g}}{4.0\ \text{g/mol}} = 0.50\ \text{mol}$$

At $T_i = 100°C = 373$ K and $p_i = 1.0$ atm $= 1.013 \times 10^5$ Pa, the gas has a volume

$$V_i = \frac{nRT_i}{p_i} = 0.0153\ \text{m}^3 = 15.3\ \text{L}$$

For an isothermal process ($T_f = T_i$) that doubles the volume $V_f = 2V_i$,

$$p_f V_f = p_i V_i \Rightarrow p_f = p_i (V_i/V_f) = (1.0\ \text{atm})\left(\tfrac{1}{2}\right) = 0.50\ \text{atm}$$

**(b)** The work done by the environment on the gas is

$$W = -nRT_i \ln(V_f/V_i) = -(0.50\ \text{mol})(8.31\ \text{J/mol K})(373\ \text{K})\ln(2) = -1074\ \text{J} \approx -1070\ \text{J}$$

**(c)** Because $\Delta E_{th} = Q + W = 0$ J for an isothermal process, the heat input to the gas is $Q = -W = 1074\ \text{J} \approx 1070\ \text{J}$.
**(d)** The change in internal energy $\Delta E_{th} = 0$ J.
**(e)**

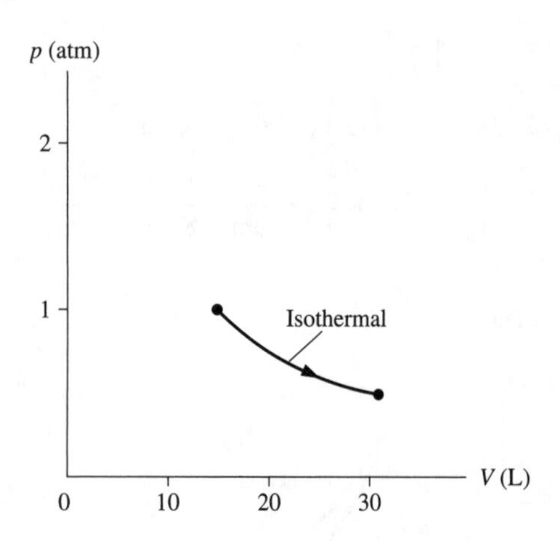


**17.71. Model:** The gas is assumed to be an ideal gas that is subjected to an isochoric process.
**Solve:** **(a)** The number of moles in 14.0 g of $N_2$ gas is

$$n = \frac{M}{M_{mol}} = \frac{14.0\ \text{g}}{28\ \text{g/mol}} = 0.50\ \text{mol}$$

At $T_i = 273$ K and $p_i = 1.0$ atm $= 1.013 \times 10^5$ Pa, the gas has a volume

$$V_i = \frac{nRT_i}{p_i} = 0.0112\ \text{m}^3 = 11.2\ \text{L}$$

For an isochoric process ($V_i = V_f$),

$$\frac{T_f}{T_i} = \frac{p_f}{p_i} = \frac{20\ \text{atm}}{1\ \text{atm}} = 20 \Rightarrow T_f = 20(273\ \text{K}) = 5460\ \text{K} \approx 5500\ \text{K}$$

**(b)** The work done on the gas is $W = -p\Delta V = 0$ J.

**(c)** The heat input to the gas is

$$Q = nC_V(T_f - T_i) = (0.50\ \text{mol})(20.8\ \text{J/mol K})(5460\ \text{K} - 273\ \text{K}) = 5.4 \times 10^4\ \text{J}$$

**(d)** The pressure ratio is

$$\frac{p_{\text{max}}}{p_{\text{min}}} = \frac{p_f}{p_i} = \frac{20\ \text{atm}}{1\ \text{atm}} = 20$$

**(e)**

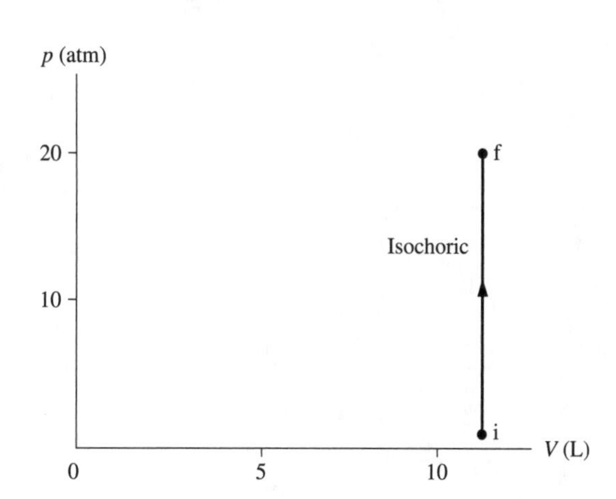

# 18

# THE MICRO/MACRO CONNECTION

**18.1.** **Solve:** We can use the ideal-gas law in the form $pV = Nk_BT$ to determine the Loschmidt number (*N*/*V*):

$$\frac{N}{V} = \frac{p}{k_BT} = \frac{(1.013\times10^5 \text{ Pa})}{(1.38\times10^{-23} \text{ J/K})(273 \text{ K})} = 2.69\times10^{25} \text{ m}^{-3}$$

**18.7.** **Solve:** The number density of the Ping-Pong balls inside the box is

$$\frac{N}{V} = \frac{2000}{1.0 \text{ m}^3} = 2000 \text{ m}^{-3}$$

With $r = (3.0 \text{ cm})/2 = 1.5$ cm, the mean free path of the balls is

$$\lambda = \frac{1}{4\sqrt{2}\pi(N/V)(r^2)} = 0.125 \text{ m} = 12.5 \text{ cm}$$

**18.11.** **Solve:** **(a)** The atomic mass number of argon is 40. This means the mass of an argon atom is

$$m = 40 \text{ u} = 40(1.661\times10^{-27} \text{ kg}) = 6.64\times10^{-26} \text{ kg}$$

The pressure of the gas is

$$p = \tfrac{1}{3}\left(\frac{N}{V}\right)mv_{\text{rms}}^2 = \tfrac{1}{3}(2.00\times10^{25} \text{ m}^{-3})(6.64\times10^{-26} \text{ kg})(455 \text{ m/s})^2 = 9.16\times10^4 \text{ Pa}$$

**(b)** The temperature of the gas in the container can be obtained from the ideal-gas equation in the form $pV = Nk_BT$:

$$T = \frac{pV}{Nk_B} = \frac{9.16\times10^4 \text{ Pa}}{(2.00\times10^{25} \text{ m}^{-3})(1.38\times10^{-23} \text{ J/K})} = 332 \text{ K}$$

**18.17.** **Solve:** The average translational kinetic energy per molecule is

$$\epsilon_{\text{avg}} = \frac{1}{2}mv_{\text{rms}}^2 = \frac{3}{2}k_BT \Rightarrow v_{\text{rms}} = \sqrt{\frac{3k_BT}{m}}$$

Since we want the $v_{\text{rms}}$ for $H_2$ and $N_2$ to be equal,

$$\sqrt{\frac{3k_BT_{H_2}}{m_{H_2}}} = \sqrt{\frac{3k_BT}{m_{N_2}}} \Rightarrow T_{H_2} = \frac{m_{H_2}}{m_{N_2}}T_{N_2} = \left(\frac{2 \text{ u}}{28 \text{ u}}\right)(373 \text{ K}) = 27 \text{ K} = -246°\text{C}$$

**18.23.** **Solve:** **(a)** The total translational kinetic energy of a gas is $K_{\text{micro}} = \frac{3}{2}N_Ak_BT = \frac{3}{2}nRT$. For $H_2$ gas at STP,

$$K_{\text{micro}} = \frac{3}{2}(1.0 \text{ mol})(8.31 \text{ J/mol K})(273 \text{ K}) = 3400 \text{ J}$$

**(b)** For He gas at STP,

$$K_{\text{micro}} = \frac{3}{2}(1.0\text{ mol})(8.31\text{ J/mol K})(273\text{ K}) = 3400\text{ J}$$

**(c)** For $O_2$ gas at STP, $K_{\text{micro}} = 3400$ J.
**Assess:** The translational kinetic energy of a gas depends on the temperature and the number of molecules but not on the molecule's mass.

**18.29. Solve:** The volume of the air is $V = 6.0\text{ m} \times 8.0\text{ m} \times 3.0\text{ m} = 144.0\text{ m}^3$, the pressure $p = 1\text{ atm} = 1.013 \times 10^5$ Pa, and the temperature $T = 20°\text{C} = 293$ K. The number of moles of the gas is

$$n = \frac{pV}{RT} = 5991\text{ mol}$$

This means the number of molecules is

$$N = nN_A = (5991\text{ mols})(6.022 \times 10^{23}\text{ mol}^{-1}) = 3.61 \times 10^{27}\text{ molecules}$$

Since air is a diatomic gas, the room's thermal energy is

$$E_{\text{th}} = N\epsilon_{\text{avg}} = N\left(\tfrac{5}{2}k_B T\right) = 3.6 \times 10^7\text{ J}$$

**Assess:** The room's thermal energy can also be obtained as follows:

$$E_{\text{th}} = nC_V T = (5991\text{ mol})(20.8\text{ J/mol K})(293\text{ K}) = 3.6 \times 10^7\text{ J}$$

**18.33. Visualize:** Refer to Figure 18.13. At low temperatures, $C_V = \frac{3}{2}R = 12.5$ J/mol K. At room temperature and modestly hot temperatures, $C_V = \frac{5}{2}R = 20.8$ J/mol K. At very hot temperatures, $C_V = \frac{7}{2}R = 29.1$ J/mol K.
**Solve:** **(a)** The number of moles of diatomic hydrogen gas in the rigid container is

$$\frac{0.20\text{ g}}{2\text{ g/mol}} = 0.10\text{ mol}$$

The heat needed to change the temperature of the gas from 50 K to 100 K at constant volume is

$$Q = \Delta E_{\text{th}} = nC_V\Delta T = (0.10\text{ mol})(12.5\text{ J/mol K})(100\text{ K} - 50\text{ K}) = 62\text{ J}$$

**(b)** To raise the temperature from 250 K to 300 K,

$$Q = \Delta E_{\text{th}} = (0.10\text{ mol})(20.8\text{ J/mol K})(300\text{ K} - 250\text{ K}) = 100\text{ J}$$

**(c)** To raise the temperature from 550 K to 600 K, $Q = 100$ J.
**(d)** To raise the temperature from 2250 K to 2300 K, $Q = \Delta E_{\text{th}} = nC_V\Delta T = (0.10\text{ mol})(29.1\text{ J/mol K})(50\text{ K}) = 150$ J.

**18.35. Solve:** **(a)** The thermal energy of a monatomic gas is

$$E_{\text{th}} = \frac{3}{2}Nk_B T = \frac{3}{2}nRT \Rightarrow T = \frac{2}{3}\frac{E_{\text{th}}}{n}\frac{1}{R}$$

$$\Rightarrow T_A = \left(\frac{2}{3}\right)\left(\frac{5000\text{ J}}{2.0\text{ mol}}\right)\frac{1}{(8.31\text{ J/mol K})} = 201\text{ K}$$

$$T_B = \left(\frac{2}{3}\right)\left(\frac{8000\text{ J}}{3.0\text{ mol}}\right)\frac{1}{(8.31\text{ J/mol K})} = 214\text{ K}$$

Thus, gas B has the higher initial temperature.
**(b)** The equilibrium condition is $(\varepsilon_A)_{\text{avg}} = (\varepsilon_B)_{\text{avg}} = (\varepsilon_{\text{tot}})_{\text{avg}}$. This means

$$\frac{E_{\text{Af}}}{n_A} = \frac{E_{\text{Bf}}}{n_B} = \frac{E_{\text{tot}}}{n_A + n_B}$$

$$\Rightarrow E_{\text{Af}} = \frac{n_A}{n_A + n_B}E_{\text{tot}} = \left(\frac{2\text{ mols}}{2\text{ mols} + 3\text{ mols}}\right)(5000\text{ J} + 8000\text{ J}) = 5200\text{ J}$$

$$E_{\text{Bf}} = \frac{n_B}{n_A + n_B}E_{\text{tot}} = \left(\frac{3\text{ mols}}{5\text{ mols}}\right)(13{,}000\text{ J}) = 7800\text{ J}$$

**18.39. Solve:** **(a)** To identify the gas, we need to determine its atomic mass number $A$ or, equivalently, the mass $m$ of each atom or molecule. The mass density $\rho$ and the number density $(N/V)$ are related by $\rho = m(N/V)$, so the mass is $m = \rho(V/N)$. From the ideal-gas law, the number density is

$$\frac{N}{V} = \frac{p}{kT} = \frac{50{,}000 \text{ Pa}}{(1.38\times10^{-23} \text{ J/K})(300 \text{ K})} = 1.208\times10^{25} \text{ m}^{-3}$$

Thus, the mass of an atom is

$$m = \rho\frac{V}{N} = \frac{8.02\times10^{-2} \text{ kg/m}^3}{1.208\times10^{25} \text{ m}^{-3}} = 6.64\times10^{-27} \text{ kg}$$

Converting to atomic mass units,

$$A = 6.64\times10^{-27} \text{ kg}\times\frac{1 \text{ u}}{1.661\times10^{-27} \text{ kg}} = 4.00 \text{ u}$$

This is the atomic mass of helium.
**(b)** Knowing the mass, we find $v_{\text{rms}}$ to be

$$v_{\text{rms}} = \sqrt{\frac{3k_BT}{m}} = \sqrt{\frac{3(1.38\times10^{-23} \text{ J/K})(300 \text{ K})}{6.64\times10^{-27} \text{ kg}}} = 1370 \text{ m/s}$$

**(c)** A typical atomic radius is $r \approx 0.5 \times 10^{-10}$ m. The mean free path is thus

$$\lambda = \frac{1}{4\sqrt{2}\pi(N/V)r^2} = \frac{1}{4\sqrt{2}\pi(1.208\times10^{25} \text{ m}^{-3})(0.5\times10^{-10} \text{ m})^2} = 1.86\times10^{-6} \text{ m} = 1.86\ \mu\text{m}$$

**18.45. Solve:** **(a)** The cylinder volume is $V = \pi r^2 L = 1.571 \times 10^{-3}$ m$^3$. Thus the number density is

$$\frac{N}{V} = \frac{2.0\times10^{22}}{1.571\times10^{-3} \text{ m}^3} = 1.273\times10^{25} \text{ m}^{-3} \approx 1.3\times10^{25} \text{ m}^{-3}$$

**(b)** The mass of an argon atom is

$$m = 40 \text{ u} = 40(1.661 \times 10^{-27} \text{ kg}) = 6.64 \times 10^{-26} \text{ kg}$$

$$\Rightarrow v_{\text{rms}} = \sqrt{\frac{3k_BT}{m}} = \sqrt{\frac{3(1.38\times10^{-23} \text{ J/K})(323 \text{ K})}{6.64\times10^{-26} \text{ kg}}} = 449 \text{ m/s} \approx 450 \text{ m/s}$$

**(c)** $v_{\text{rms}}$ is the square root of the average of $v^2$. That is,

$$v_{\text{rms}}^2 = (v^2)_{\text{avg}} = (v_x^2)_{\text{avg}} + (v_y^2)_{\text{avg}} + (v_z^2)_{\text{avg}}$$

An atom is equally likely to move in the $x$, $y$, or $z$ direction, so *on average* $(v_x^2)_{\text{avg}} = (v_y^2)_{\text{avg}} = (v_z^2)_{\text{avg}}$. Hence,

$$v_{\text{rms}}^2 = 3(v_x^2)_{\text{avg}} \Rightarrow (v_x)_{\text{rms}} = \sqrt{(v_x^2)_{\text{avg}}} = \frac{v_{\text{rms}}}{\sqrt{3}} = 259 \text{ m/s} \approx 260 \text{ m/s}$$

**(d)** When we considered all the atoms to have the same velocity, we found the collision rate to be $\frac{1}{2}(N/V)Av_x$ (see Equation 18.10). Because the atoms move with different speeds, we need to replace $v_x$ with $(v_x)_{\text{rms}}$. The end of the cylinder has area $A = \pi r^2 = 7.85 \times 10^{-3}$ m$^2$. Therefore, the number of collisions per second is

$$\tfrac{1}{2}(N/V)A(v_x)_{\text{rms}} = \tfrac{1}{2}(1.273\times10^{25} \text{ m}^{-3})(7.85\times10^{-3} \text{ m}^2)(259 \text{ m/s}) = 1.3 \times 10^{25} \text{ s}^{-1}$$

**(e)** From kinetic theory, the pressure is

$$p = \tfrac{1}{3}\left(\frac{N}{V}\right)m(v^2)_{\text{avg}} = \tfrac{1}{3}\left(\frac{N}{V}\right)mv_{\text{rms}}^2 = \tfrac{1}{3}(1.273\times10^{25} \text{ m}^{-3})(6.64\times10^{-26} \text{ kg})(449 \text{ m/s})^2 = 56{,}800 \text{ Pa} \approx 57{,}000 \text{ Pa}$$

**(f)** From the ideal-gas law, the pressure is

$$p = \frac{Nk_BT}{V} = \frac{(2.0\times10^{22})(1.38\times10^{-23}\ \text{J/K})(323\ \text{K})}{1.571\times10^{-3}\ \text{m}^3} = 56{,}700\ \text{Pa} \approx 57{,}000\ \text{Pa}$$

**Assess:** The very slight difference between parts (e) and (f) is due to rounding errors; to two significant figures they are the same.

**18.49. Solve:** **(a)** The number of moles of helium and oxygen are

$$n_{\text{helium}} = \frac{2.0\ \text{g}}{4.0\ \text{g/mol}} = 0.50\ \text{mol} \qquad n_{\text{oxygen}} = \frac{8.0\ \text{g}}{32.0\ \text{g/mol}} = 0.25\ \text{mol}$$

Since helium is a monoatomic gas, the initial thermal energy is

$$E_{\text{helium i}} = n_{\text{helium}}\left(\tfrac{3}{2}RT_{\text{helium}}\right) = (0.50\ \text{mol})\left(\tfrac{3}{2}\right)(8.31\ \text{J/mol K})(300\ \text{K}) = 1870\ \text{J} \approx 1900\ \text{J}$$

Since oxygen is a diatomic gas, the initial thermal energy is

$$E_{\text{oxygen i}} = n_{\text{oxygen}}\left(\tfrac{5}{2}RT_{\text{oxygen}}\right) = (0.25\ \text{mol})\left(\tfrac{5}{2}\right)(8.31\ \text{J/mol K})(600\ \text{K}) = 3116\ \text{J} \approx 3100\ \text{J}$$

**(b)** The total initial thermal energy is

$$E_{\text{tot}} = E_{\text{helium i}} + E_{\text{oxygen i}} = 4986\ \text{J}$$

As the gases interact, they come to equilibrium at a common temperature $T_f$. This means

$$4986\ \text{J} = n_{\text{helium}}\left(\tfrac{3}{2}RT_f\right) + n_{\text{oxygen}}\left(\tfrac{5}{2}RT_f\right)$$

$$\Rightarrow T_f = \frac{4986\ \text{J}}{\left(\tfrac{1}{2}R\right)(3n_{\text{helium}} + 5n_{\text{oxygen}})} = \frac{4986\ \text{J}}{\tfrac{1}{2}(8.31\ \text{J/mol K})(3\times0.50\ \text{mol} + 5\times0.25\ \text{mol})} = 436.4\ \text{K} = 436\ \text{K}$$

The thermal energies at the final temperature $T_f$ are

$$E_{\text{helium f}} = n_{\text{helium}}\left(\tfrac{3}{2}RT_f\right) = \left(\tfrac{3}{2}\right)(0.50\ \text{mol})(8.31\ \text{J/mol K})(436.4\ \text{K}) = 2700\ \text{J}$$

$$E_{\text{oxygen f}} = n_{\text{oxygen}}\left(\tfrac{5}{2}RT_f\right) = \left(\tfrac{5}{2}\right)(0.25\ \text{mol})(8.31\ \text{J/mol K})(436.4\ \text{K}) = 2300\ \text{J}$$

**(c)** The change in the thermal energies are

$$E_{\text{helium f}} - E_{\text{helium i}} = 2720\ \text{J} - 1870\ \text{J} = 850\ \text{J} \qquad E_{\text{oxygen f}} - E_{\text{oxygen i}} = 2266\ \text{J} - 3116\ \text{J} = -850\ \text{J}$$

The helium gains energy and the oxygen loses energy.
**(d)** The final temperature can also be calculated as follows:

$$E_{\text{helium f}} = (n_{\text{helium}})\tfrac{3}{2}RT_f \Rightarrow 2720\ \text{J} = (0.50\ \text{mol})(1.5)(8.31\ \text{J/mol K})T_f \Rightarrow T_f = 436.4\ \text{K} \approx 436\ \text{K}$$

**18.55. Model:** Assume the gas is monatomic.
**Visualize:** From the equipartition theorem there is $\tfrac{1}{2}nRT$ of energy for each degree of freedom. For a two-dimensional monatomic gas there are only two degrees of freedom.
**Solve:** **(a)** $C_V = \frac{2}{2}R = R$
**(b)** Equation 17.34 gives $C_p = C_V + R$, so

$$C_p = R + R = 2R$$

**Assess:** It would be nice to measure the values and compare with these predictions.

**18.57. Solve:** **(a)** The rms speed is

$$v_{\text{rms}} = \sqrt{\frac{3k_BT}{m}} \Rightarrow \frac{v_{\text{rms hydrogen}}}{v_{\text{rms oxygen}}} = \sqrt{\frac{32\ \text{u}}{2\ \text{u}}} = 4$$

**(b)** The average translational energy is $\epsilon = \frac{3}{2}k_B T$. Thus

$$\frac{\epsilon_{\text{avg hydrogen}}}{\epsilon_{\text{avg oxygen}}} = \frac{T_{\text{hydrogen}}}{T_{\text{oxygen}}} = 1$$

**(c)** The thermal energy is

$$E_{\text{th}} = \tfrac{5}{2}nRT$$

$$\Rightarrow \frac{E_{\text{th hydrogen}}}{E_{\text{th oxygen}}} = \frac{n_{\text{hydrogen}}}{n_{\text{oxygen}}} = \frac{m_{\text{hydrogen}}}{2.0\text{ g/mol}}\frac{32.0\text{ g/mol}}{m_{\text{oxygen}}} = 16$$

**18.61. Solve:** **(a)** The thermal energy is

$$E_{\text{th}} = (E_{\text{th}})_{N_2} + (E_{\text{th}})_{O_2} = \tfrac{5}{2}N_{N_2}k_B T + \tfrac{5}{2}N_{O_2}k_B T = \tfrac{5}{2}N_{\text{total}}k_B T$$

where $N_{\text{total}}$ is the total number of molecules. The identity of the molecules makes no difference since both are diatomic. The number of molecules in the room is

$$N_{\text{total}} = \frac{pV}{k_B T} = \frac{(101{,}300\text{ Pa})(2\text{ m}\times 2\text{ m}\times 2\text{ m})}{(1.38\times 10^{-23}\text{ J/K})(273\text{ K})} = 2.15\times 10^{26}$$

The thermal energy is

$$E_{\text{th}} = \tfrac{5}{2}(2.15\times 10^{26})(1.38\times 10^{-23}\text{ J/K})(273\text{ K}) = 2.03\times 10^{6}\text{ J} \approx 2.0\times 10^{6}\text{ J}$$

**(b)** A 1 kg ball at height $y = 1$ m has a potential energy $U = mgy = 9.8$ J. The ball would need 9.8 J of initial kinetic energy to reach this height. The fraction of thermal energy that would have to be conveyed to the ball is

$$\frac{9.8\text{ J}}{2.03\times 10^{6}\text{ J}} = 4.8\times 10^{-6}$$

**(c)** A temperature change $\Delta T$ corresponds to a thermal energy change $\Delta E_{\text{th}} = \frac{5}{2}N_{\text{total}}k_B\Delta T$. But $\frac{5}{2}N_{\text{total}}k_B = E_{\text{th}}/T$. Using this, we can write

$$\Delta E_{\text{th}} = \frac{E_{\text{th}}}{T}\Delta T \Rightarrow \Delta T = \frac{\Delta E_{\text{th}}}{E_{\text{th}}}T = \frac{-9.8\text{ J}}{2.03\times 10^{6}\text{ J}}273\text{ K} = -0.0013\text{ K}$$

The room temperature would decrease by 0.0013 K or 0.0013°C.

**(d)** The situation with the ball at rest on the floor and in thermal equilibrium with the air is a very probable distribution of energy and thus a state with high entropy. Although energy would be conserved by removing energy from the air and transferring it to the ball, this would be a very *improbable* distribution of energy and thus a state of low entropy. The ball will not be spontaneously launched from the ground because this would require a decrease in entropy, in violation of the second law of thermodynamics.

As another way of thinking about the situation, the ball and the air are initially at the same temperature. Once even the slightest amount of energy is transferred from the air to the ball, the air's temperature will be less than that of the ball. Any further flow of energy from the air to the ball would be a situation in which heat energy is flowing from a colder object to a hotter object. This cannot happen because it would violate the second law of thermodynamics.

# 19

## HEAT ENGINES AND REFRIGERATORS

**19.1. Solve:** **(a)** The engine has a thermal efficiency of $\eta = 40\% = 0.40$ and a work output of 100 J per cycle. The heat input is calculated as follows:

$$\eta = \frac{W_{\text{out}}}{Q_{\text{H}}} \Rightarrow 0.40 = \frac{100\text{ J}}{Q_{\text{H}}} \Rightarrow Q_{\text{H}} = 250\text{ J}$$

**(b)** Because $W_{\text{out}} = Q_{\text{H}} - Q_{\text{C}}$, the heat exhausted is

$$Q_{\text{C}} = Q_{\text{H}} - W_{\text{out}} = 250\text{ J} - 100\text{ J} = 150\text{ J}$$

**19.7. Solve:** The amount of heat discharged per second is calculated as follows:

$$\eta = \frac{W_{\text{out}}}{Q_{\text{H}}} = \frac{W_{\text{out}}}{Q_{\text{C}} + W_{\text{out}}} \Rightarrow Q_{\text{C}} = W_{\text{out}}\left(\frac{1}{\eta} - 1\right) = (900\text{ MW})\left(\frac{1}{0.32} - 1\right) = 1.913\times10^{9}\text{ W}$$

That is, each second the electric power plant discharges $1.913\times10^{9}$ J of energy into the ocean. Since a typical American house needs $2.0\times10^{4}$ J of energy per second for heating, the number of houses that could be heated with the waste heat is $(1.913\times10^{9}\text{ J})/(2.0\times10^{4}\text{ J}) = 96{,}000$.

**19.9. Model:** Process A is isochoric, process B is isothermal, process C is adiabatic, and process D is isobaric.
**Solve:** Process A is isochoric, so the increase in pressure increases the temperature and hence the thermal energy. Because $\Delta E_{\text{th}} = Q - W_{\text{s}}$ and $W_{\text{s}} = 0$ J, $Q$ increases for process A. Process B is isothermal, so $T$ is constant and hence $\Delta E_{\text{th}} = 0$ J. The work done $W_{\text{s}}$ is positive because the gas expands. Because $Q = W_{\text{s}} + \Delta E_{\text{th}}$, $Q$ is positive for process B. Process C is adiabatic, so $Q = 0$ J. $W_{\text{s}}$ is positive because of the increase in volume. Since $Q = 0\text{ J} = W_{\text{s}} + \Delta E_{\text{th}}$, $\Delta E_{\text{th}}$ is negative for process C. Process D is isobaric, so the decrease in volume leads to a decrease in temperature and hence a decrease in the thermal energy. Due to the decrease in volume, $W_{\text{s}}$ is negative. Because $Q = W_{\text{s}} + \Delta E_{\text{th}}$, $Q$ also decreases for process D.

| | $\Delta E_{\text{th}}$ | $W_{\text{s}}$ | $Q$ |
|---|---|---|---|
| A | + | 0 | + |
| B | 0 | + | + |
| C | – | + | 0 |
| D | – | – | – |

**19.13. Model:** The heat engine follows a closed cycle, starting and ending in the original state. The cycle consists of three individual processes.
**Solve:** **(a)** The work done by the heat engine per cycle is the area enclosed by the $p$-versus-$V$ graph. We get

$$W_{\text{out}} = \tfrac{1}{2}(200\text{ kPa})(100\times10^{-6}\text{ m}^3) = 10\text{ J}$$

The heat energy transferred into the engine is $Q_{\text{H}} = 30\text{ J} + 84\text{ J} = 114\text{ J}$. Because $W_{\text{out}} = Q_H - Q_C$, the heat energy exhausted is

$$Q_C = Q_H - W_{out} = 114\text{ J} - 10\text{ J} = 104\text{ J}$$

**(b)** The thermal efficiency of the engine is

$$\eta = \frac{W_{out}}{Q_H} = \frac{10\text{ J}}{114\text{ J}} = 0.088$$

**Assess:** Practical engines have thermal efficiencies in the range $\eta \approx 0.1 - 0.4$.

**19.17. Model:** The Brayton cycle involves two adiabatic processes and two isobaric processes.
**Solve:** From Equation 19.21, the efficiency of a Brayton cycle is $\eta_B = 1 - r_p^{(1-\gamma)/\gamma}$, where $r_p$ is the pressure ratio $p_{max}/p_{min}$. The specific heat ratio for a diatomic gas is

$$\gamma = \frac{C_P}{C_V} = \frac{\frac{7}{2}\text{R}}{\frac{5}{2}\text{R}} = 1.4$$

Solving the above equation for $r_p$,

$$(1-\eta_B) = r_p^{(1-\gamma)/\gamma} \Rightarrow r_p = (1-\eta_B)^{\frac{\gamma}{1-\gamma}} = (1-0.60)^{\frac{-1.4}{0.4}} = 25$$

**19.19. Model:** The efficiency of a Carnot engine ($\eta_{Carnot}$) depends only on the temperatures of the hot and cold reservoirs. On the other hand, the thermal efficiency ($\eta$) of a heat engine depends on the heats $Q_H$ and $Q_C$.
**Solve:** **(a)** According to the first law of thermodynamics, $Q_H = W_{out} + Q_C$. For engine (a), $Q_H = 50$ J, $Q_C = 20$ J and $W_{out} = 30$ J, so the first law of thermodynamics is obeyed. For engine (b), $Q_H = 10$ J, $Q_C = 7$ J and $W_{out} = 4$ J, so the first law is violated. For engine (c) the first law of thermodynamics is obeyed.
**(b)** For the three heat engines, the maximum or Carnot efficiency is

$$\eta_{Carnot} = 1 - \frac{T_C}{T_H} = 1 - \frac{300\text{ K}}{600\text{ K}} = 0.50$$

Engine (a) has

$$\eta = 1 - \frac{Q_C}{Q_H} = \frac{W_{out}}{Q_H} = \frac{30\text{ J}}{50\text{ J}} = 0.60$$

This is larger than $\eta_{Carnot}$, thus violating the second law of thermodynamics. For engine (b),

$$\eta = \frac{W_{out}}{Q_H} = \frac{4\text{ J}}{10\text{ J}} = 0.40 < \eta_{Carnot}$$

so the second law is obeyed. Engine (c) has a thermal efficiency that is

$$\eta = \frac{10\text{ J}}{30\text{ J}} = 0.333 < \eta_{Carnot}$$

so the second law of thermodynamics is obeyed.

**19.23. Model:** The efficiency of an ideal engine (or Carnot engine) depends only on the temperatures of the hot and cold reservoirs.
**Solve:** **(a)** The engine's thermal efficiency is

$$\eta = \frac{W_{out}}{Q_H} = \frac{W_{out}}{Q_C + W_{out}} = \frac{10\text{ J}}{15\text{ J} + 10\text{ J}} = 0.40 = 40\%$$

**(b)** The efficiency of a Carnot engine is $\eta_{Carnot} = 1 - T_C/T_H$. The minimum temperature in the hot reservoir is found as follows:

$$0.40 = 1 - \frac{293\text{ K}}{T_H} \Rightarrow T_H = 488\text{ K} = 215°\text{C}$$

This is the minimum possible temperature. In a real engine, the hot-reservoir temperature would be higher than 215°C because no real engine can match the Carnot efficiency.

**19.29. Model:** The minimum possible value of $T_C$ occurs with a Carnot refrigerator.
**Solve:** **(a)** For the refrigerator, the coefficient of performance is

$$K = \frac{Q_C}{W_{in}} \Rightarrow Q_C = KW_{in} = (5.0)(10\text{ J}) = 50\text{ J}$$

The heat energy exhausted per cycle is

$$Q_H = Q_C + W_{in} = 50\text{ J} + 10\text{ J} = 60\text{ J}$$

**(b)** If the hot-reservoir temperature is 27°C = 300 K, the lowest possible temperature of the cold reservoir can be obtained as follows:

$$K_{Carnot} = \frac{T_C}{T_H - T_C} \Rightarrow 5.0 = \frac{T_C}{300\text{ K} - T_C} \Rightarrow T_C = 250\text{ K} = -23°\text{C}$$

**19.37. Visualize:** We are given $T_H = 298$ K and $T_C = 273$ K . See Figure 19.11.

**Solve:** $Q_C = mL_f = (10\text{kg})(3.33\times10^5\text{ J/kg}) = 3.33\times10^6\text{ J}$ .

**(a)** For a Carnot cycle $\eta_{Carnot} = 1 - \frac{T_C}{T_H}$ but that must also equal $\eta = 1 - \frac{Q_C}{Q_H}$, so $\frac{Q_C}{Q_H} = \frac{T_C}{T_H}$.

$$Q_H = Q_C\frac{T_H}{T_C} = (3.33\times10^6\text{ J})\left(\frac{298\text{K}}{273\text{K}}\right) = 3.63\times10^6\text{ J}$$

**(b)**

$$W_{in} = Q_H - Q_C = 3.63\times10^6\text{ J} - 3.33\times10^6\text{ J} = 0.30\times10^6\text{ J} = 3.0\times10^5\text{ J}$$

**Assess:** This is a reasonable amount of work to freeze 10 kg of water.

**19.45. Visualize:** We are given $T_C = 275$ K , $T_H = 295$ K. We are also given that in one second $W_{in} = 100$ J and $Q_C = (1\text{ s})(100\text{ kJ/min})(1\text{ min}/60\text{ s}) = 1667$ J.
**Solve:** The coefficient of performance of a refrigerator is given in Equation 19.10.

$$K = \frac{Q_C}{W_{in}} = \frac{1667\text{ J}}{100\text{ J}} = 16.67$$

However the coefficient of performance of a Carnot refrigerator is given in Equation 19.30.

$$K_{Carnot} = \frac{T_C}{T_H - T_C} = \frac{275\text{ K}}{20\text{ K}} = 13.75$$

However, informal statement #8 of the second law says that the coefficient of performance cannot exceed the Carnot coefficient of performance, so the salesman is making false claims. You should not buy the DreamFridge.
**Assess:** The second law imposes real-world restrictions.

**19.47. Model:** The power plant is to be treated as a heat engine.
**Solve:** **(a)** Every hour 300 metric tons or 3 x $10^5$ kg of coal is burnt. The volume of coal is

$$\frac{3\times10^5\text{ kg}}{1\text{ h}}\times\frac{\text{m}^3}{1500\text{ kg}}\times24\text{ h} = 4800\text{ m}^3$$

The height of the room will be 48 m.
**(b)** The thermal efficiency of the power plant is

$$\eta = \frac{W_{out}}{Q_H} = \frac{7.50\times10^8\text{ J/s}}{\frac{3\times10^5\text{ kg}}{1\text{ h}}\times\frac{28\times10^6\text{ J}}{\text{kg}}\times\frac{1\text{ h}}{3600\text{ s}}} = \frac{7.50\times10^8\text{ J}}{2.333\times10^9\text{ J}} = 0.32 = 32\%$$

**Assess:** An efficiency of 32% is typical of power plants.

**19.53. Model:** The heat engine follows a closed cycle. For a diatomic gas, $C_V = \frac{5}{2}R$ and $C_P = \frac{7}{2}R$.

**Visualize:** Please refer to Figure P19.53.

**Solve:** (a) Since $T_1 = 293$ K, the number of moles of the gas is

$$n = \frac{p_1V_1}{RT_1} = \frac{(0.5\times1.013\times10^5 \text{ Pa})(10\times10^{-6} \text{ m}^3)}{(8.31 \text{ J/mol K})(293 \text{ K})} = 2.08\times10^{-4} \text{ mol}$$

At point 2, $V_2 = 4V_1$ and $p_2 = 3p_1$. The temperature is calculated as follows:

$$\frac{p_1V_1}{T_1} = \frac{p_2V_2}{T_2} \Rightarrow T_2 = \frac{p_2}{p_1}\frac{V_2}{V_1}T_1 = (3)(4)(293 \text{ K}) = 3516 \text{ K}$$

At point 3, $V_3 = V_2 = 4V_1$ and $p_3 = p_1$. The temperature is calculated as before:

$$T_3 = \frac{p_3}{p_1}\frac{V_3}{V_1}T_1 = (1)(4)(293 \text{ K}) = 1172 \text{ K}$$

For process 1 → 2, the work done is the area under the $p$-versus-$V$ curve. That is,

$$W_s = (0.5 \text{ atm})(40 \text{ cm}^3 - 10 \text{ cm}^3) + \tfrac{1}{2}(1.5 \text{ atm} - 0.5 \text{ atm})(40 \text{ cm}^3 - 10 \text{ cm}^3)$$
$$= (30\times10^{-6} \text{ m}^3)(1 \text{ atm})\left(\frac{1.013\times10^5 \text{ Pa}}{1 \text{ atm}}\right) = 3.04 \text{ J}$$

The change in the thermal energy is

$$\Delta E_{th} = nC_V\Delta T = (2.08\times10^{-4} \text{ mol})\tfrac{5}{2}(8.31 \text{ J/mol K})(3516 \text{ K} - 293 \text{ K}) = 13.93 \text{ J}$$

The heat is $Q = W_s + \Delta E_{th} = 16.97$ J. For process 2 → 3, the work done is $W_s = 0$ J and

$$Q = \Delta E_{th} = nC_V\Delta T = n(\tfrac{5}{2}R)(T_3 - T_2)$$
$$= (2.08\times10^{-4} \text{ mol})\tfrac{5}{2}(8.31 \text{ J/mol K})(1172 \text{ K} - 3516 \text{ K}) = -10.13 \text{ J}$$

For process 3 → 1,

$$W_s = (0.5 \text{ atm})(10 \text{ cm}^3 - 40 \text{ cm}^3) = (0.5\times1.013\times10^5 \text{ Pa})(-30\times10^{-6} \text{ m}^3) = -1.52 \text{ J}$$
$$\Delta E_{th} = nC_V\Delta T = (2.08\times10^{-4} \text{ mol})\tfrac{5}{2}(8.31 \text{ J/mol K})(293 \text{ K} - 1172 \text{ K}) = -3.80 \text{ J}$$

The heat is $Q = \Delta E_{th} + W_s = -5.32$ J.

| | $W_s$ (J) | $Q$ (J) | $\Delta E_{th}$ |
|---|---|---|---|
| 1 → 2 | 3.04 | 16.97 | 13.93 |
| 2 → 3 | 0 | -10.13 | -10.13 |
| 3 → 1 | -1.52 | -5.32 | -3.80 |
| Net | 1.52 | 1.52 | 0 |

**(b)** The efficiency of the engine is

$$\eta = \frac{W_{net}}{Q_H} = \frac{1.52 \text{ J}}{16.97 \text{ J}} = 0.090 = 9.0\%$$

**(c)** The power output of the engine is

$$500 \frac{\text{revolutions}}{\text{min}} \times \frac{1 \text{ min}}{60 \text{ s}} \times \frac{W_{net}}{\text{revolution}} = \frac{500}{60} \times 1.52 \text{ J/s} = 13 \text{ W}$$

**Assess:** For a closed cycle, as expected, $(W_s)_{net} = Q_{net}$ and $(\Delta E_{th})_{net} = 0$ J.

**19.57. Model:** For the closed cycle of the heat engine, process 1 → 2 is adiabatic, process 2 → 3 is isothermal, and process 3 → 1 is isobaric. For a diatomic gas $C_V = \frac{5}{2}R$ and $\gamma = \frac{7}{5}$.

**Solve:** (a) From the graph $V_1 = 3000 \text{cm}^3$ and $p_1 = 100$ kPa.

The number of moles of gas is

$$n = \frac{p_2V_2}{RT_2} = \frac{(4.0\times10^5\ \text{Pa})(1000\times10^{-6}\ \text{m}^3)}{(8.31\ \text{J/mol K})(400\ \text{K})} = 0.1203\ \text{mol}$$

With $p_1$, $V_1$, and $nR$ having been determined, we can find $T_1$ using the ideal-gas equation:

$$T_1 = \frac{p_1V_1}{nR} = \frac{(100\times10^3\ \text{kPa})(3000\times10^{-6}\ \text{m}^3)}{(8.31\ \text{J/mol K})(0.1203\ \text{mol})} = 300\ \text{K}$$

**(b)** For adiabatic process 1 → 2, $Q = 0$ J and

$$W_s = \frac{p_2V_2 - p_1V_1}{1-\gamma} = \frac{(4.0\times10^5\ \text{Pa})(1000\times10^{-6}\ \text{m}^3) - (1.0\times10^5\ \text{Pa})(3000\times10^{-6}\ \text{m}^3)}{1-1.4} = -250\ \text{J}$$

Because $\Delta E_{th} = -W_s + Q$, $\Delta E_{th} = -W_s = 250\ \text{J}$.

For isothermal process 2 → 3, $\Delta E_{th} = 0$ J. From Equation 17.16,

$$W_s = nRT_2 \ln\frac{V_3}{V_2} = 554\ \text{J}$$

From the first law of thermodynamics, $Q = W_s = 554\ \text{J}$ for process 2 → 3.

For isobaric process 3 → 1, $W_s$ = area under $p$-versus-$V$ graph=

$$W_s = (100\times10^3\ \text{Pa})(3000\times10^{-6}\ \text{m}^3 - 4000\times10^{-6}\ \text{m}^3) = -100\ \text{J}$$

$$Q_{3\to1} = nC_P(T_1 - T_3) = n\left(\frac{7}{2}R\right)(T_1 - T_3) = -350\ \text{J}$$

$$\Delta E_{th} = nC_V(T_1 - T_3) = n\left(\frac{5}{2}R\right)(T_1 - T_3) = (0.1203\ \text{mol})\left(\frac{5}{2}8.31\ \text{J/mol K}\right)(-100\ \text{K})$$

| | $\Delta E_{th}$ (J) | $W_s$ (J) | $Q$ (J) |
|---|---|---|---|
| 1 → 2 | 250 | –250 | 0 |
| 2 → 3 | 0 | 554 | 554 |
| 3 → 1 | –250 | –100 | –350 |
| Net | 0 | 204 | 204 |

**(c)** The work per cycle is 204 J and the thermal efficiency of the engine is

$$\eta = \frac{W_s}{Q_H} = \frac{204\ \text{J}}{554\ \text{J}} = 0.37 = 37\%$$

**Assess:** As expected, for a closed cycle $(W_s)_{net} = Q_{net}$ and $(\Delta E_{th})_{net} = 0$ J.

**19.59. Model:** Process 1 → 2 of the cycle is isochoric, process 2 → 3 is isothermal, and process 3 → 1 is isobaric. For a monatomic gas, $C_V = \frac{3}{2}R$ and $C_P = \frac{5}{2}$ R.

**Visualize:** Please refer to Figure P19.59.

**Solve:** **(a)** At point 1: The pressure $p_1 = 1.0\ \text{atm} = 1.013\times10^5\ \text{Pa}$ and the volume $V_1 = 1000\times10^{-6}\ \text{m}^3 = 1.0\times10^{-3}\ \text{m}^3$. The number of moles is

$$n = \frac{0.120\ \text{g}}{4\ \text{g/mol}} = 0.030\ \text{mol}$$

Using the ideal-gas law,

$$T_1 = \frac{p_1V_1}{nR} = \frac{(1.013\times10^5\ \text{Pa})(1.0\times10^{-3}\ \text{m}^3)}{(0.030\ \text{mol})(8.31\ \text{J/mol K})} = 406\ \text{K}$$

At point 2: The pressure $p_2 = 5.0 \text{ atm} = 5.06\times10^5 \text{ Pa}$ and $V_2 = 1.0\times10^{-3} \text{ m}^3$. The temperature is

$$T_2 = \frac{p_2V_2}{nR} = \frac{(5.06\times10^5 \text{ Pa})(1.0\times10^{-3} \text{ m}^3)}{(0.030 \text{ mol})(8.31 \text{ J/mol K})} = 2030 \text{ K}$$

At point 3: The pressure is $p_3 = 1.0 \text{ atm} = 1.013\times10^5 \text{ Pa}$ and the temperature is $T_3 = T_2 = 2030 \text{ K}$. The volume is

$$V_3 = V_2\frac{p_2}{p_3} = (1.0\times10^{-3} \text{ m}^3)\left(\frac{5 \text{ atm}}{1 \text{ atm}}\right) = 5.0\times10^{-3} \text{ m}^3$$

**(b)** For isochoric process 1 → 2, $W_{1\to2} = 0$ J and

$$Q_{1\to2} = nC_V\Delta T = (0.030 \text{ mol})(\tfrac{3}{2}R)(2030 \text{ K} - 406 \text{ K}) = 607 \text{ J}$$

For isothermal process 2 → 3, $\Delta E_{\text{th } 2\to3} = 0$ J and

$$Q_{2\to3} = W_{2\to3} = nRT_2\ln\frac{V_3}{V_2} = (0.030 \text{ mol})(8.31 \text{ J/mol K})(2030 \text{ K})\ln\left(\frac{5.0\times10^{-3} \text{ m}^3}{1.0\times10^{-3} \text{ m}^3}\right) = 815 \text{ J}$$

For isobaric process 3 → 1,

$$W_{3\to1} = p_3\Delta V = (1.013\times10^5 \text{ Pa})(1.0\times10^{-3} \text{ m}^3 - 5.0\times10^{-3} \text{ m}^3) = -405 \text{ J}$$

$$Q_{3\to1} = nC_P\Delta T = (0.030 \text{ mol})(\tfrac{5}{2})(8.31 \text{ J/mol K})(406 \text{ K} - 2030 \text{ K}) = -1012 \text{ J}$$

The total work done is $W_{net} = W_{1\to2} + W_{2\to3} + W_{3\to1}$ = 410 J. The total heat input is $Q_H = Q_{1\to2} + Q_{2\to3} = 1422 \text{ J}$. The efficiency of the engine is

$$\eta = \frac{W_{net}}{Q_H} = \frac{410 \text{ J}}{1422 \text{ J}} = 29\%$$

**(c)** The maximum possible efficiency of a heat engine that operates between $T_{max}$ and $T_{min}$ is

$$\eta_{max} = 1 - \frac{T_{min}}{T_{max}} = 1 - \frac{406 \text{ K}}{2030 \text{ K}} = 80\%$$

**Assess:** The actual efficiency of an engine is less than the maximum possible efficiency.

**19.63. Model:** The closed cycle of the heat engine involves the following four processes: isothermal expansion, isochoric cooling, isothermal compression, and isochoric heating. For a monatomic gas $C_V = \frac{3}{2}R$.

**Visualize:**

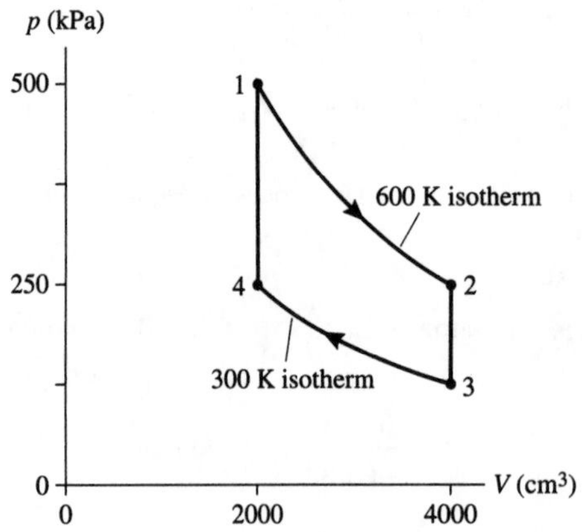

**Solve:** Using the ideal-gas law,

$$p_1 = \frac{nRT_1}{V_1} = \frac{(0.20 \text{ mol})(8.31 \text{ J/mol K})(600 \text{ K})}{2.0\times10^{-3} \text{ m}^3} = 4.986\times10^5 \text{ Pa}$$

At point 2, because of the isothermal conditions, $T_2 = T_1 = 600$ K and

$$p_2 = p_1\frac{V_1}{V_2} = \left(4.986\times10^5 \text{ Pa}\right)\left(\frac{2.0\times10^{-3} \text{ m}^3}{4.0\times10^{-3} \text{ m}^3}\right) = 2.493\times10^5 \text{ Pa}$$

At point 3, because it is an isochoric process, $V_3 = V_2 = 4000 \text{ cm}^3$ and

$$p_3 = p_2\frac{T_3}{T_2} = \left(2.493\times10^5 \text{ Pa}\right)\left(\frac{300 \text{ K}}{600 \text{ K}}\right) = 1.247\times10^5 \text{ Pa}$$

Likewise at point 4, $T_4 = T_3 = 300$ K and

$$p_4 = p_3\frac{V_3}{V_4} = \left(1.247\times10^5 \text{ Pa}\right)\left(\frac{4.0\times10^{-3} \text{ m}^3}{2.0\times10^{-3} \text{ m}^3}\right) = 2.493\times10^5 \text{ Pa}$$

Let us now calculate $W_{net} = W_{1\to2} + W_{2\to3} + W_{3\to4} + W_{4\to1}$. For the isothermal processes,

$$W_{1\to2} = nRT_1\ln\frac{V_2}{V_1} = (0.20 \text{ mol})(8.31 \text{ J/mol K})(600 \text{ K})\ln(2) = 691.2 \text{ J}$$

$$W_{3\to4} = nRT_3\ln\frac{V_4}{V_3} = (0.20 \text{ mol})(8.31 \text{ J/mol K})(300 \text{ K})\ln\left(\tfrac{1}{2}\right) = -345.6 \text{ J}$$

For the isochoric processes, $W_{2\to3} = W_{4\to1} = 0$ J. Thus, $W_{net} = 345.6 \text{ J} \approx 350 \text{ J}$.
Because $Q = W_s + \Delta E_{th}$,

$$Q_{1\to2} = W_{1\to2} + (\Delta E_{th})_{1\to2} = 691.2 \text{ J} + 0 \text{ J} = 691.2 \text{ J}$$

For the first isochoric process,

$$Q_{2\to3} = nC_V\Delta T = (0.20 \text{ mol})\left(\tfrac{3}{2}R\right)(T_3 - T_2)$$
$$= (0.20 \text{ mol})\tfrac{3}{2}(8.31 \text{ J/mol K})(300 \text{ K} - 600 \text{ K}) = -747.9 \text{ K}$$

For the second isothermal process

$$Q_{3\to4} = W_{3\to4} + (\Delta E_{th})_{3\to4} = -345.6 \text{ J} + 0 \text{ J} = -345.6 \text{ J}$$

For the second isochoric process,

$$Q_{4\to1} = nC_V\Delta T = n\left(\tfrac{3}{2}R\right)(T_1 - T_4)$$
$$= (0.20 \text{ mol})\left(\tfrac{3}{2}\right)(8.31 \text{ J/mol K})(600 \text{ K} - 300 \text{ K}) = 747.9 \text{ K}$$

Thus, $Q_H = Q_{1\to2} + Q_{4\to1} = 1439.1$ J. The thermal efficiency of the engine is

$$\eta = \frac{W_{net}}{Q_H} = \frac{345.6 \text{ J}}{1439.1 \text{ J}} = 0.24 = 24\%$$

**19.67. Solve:** **(a)** A heat engine operates at 20% efficiency and produces 20 J of work in each cycle. What is the net heat extracted from the hot reservoir and the net heat exhausted in each cycle?
**(b)** We have $0.20 = 1 - Q_C/Q_H$. Using the first law of thermodynamics,

$$W_{out} = Q_H - Q_C = 20 \text{ J} \Rightarrow Q_C = Q_H - 20 \text{ J}$$

Substituting into the definition of efficiency,

$$0.20 = 1 - \frac{Q_H - 20 \text{ J}}{Q_H} = 1 - 1 + \frac{20 \text{ J}}{Q_H} = \frac{20 \text{ J}}{Q_H} \Rightarrow Q_H = \frac{20 \text{ J}}{0.20} = 100 \text{ J}$$

The heat exhausted is $Q_C = Q_H - 20 \text{ J} = 100 \text{ J} - 20 \text{ J} = 80 \text{ J}$.